Kenzan Method for Scaffold-Free Biofabrication

Koichi Nakayama

Editor

Kenzan Method for Scaffold-Free Biofabrication

 Springer

Editor
Koichi Nakayama
Center of Regenerative Medicine Research
Faculty of Medicine
Saga University
Saga, Japan

ISBN 978-3-030-58687-4   ISBN 978-3-030-58688-1   (eBook)
https://doi.org/10.1007/978-3-030-58688-1

This Springer imprint is published by the registered company Springer Nature Switzerland AG
The registered company address is: Gewerbestrasse 11, 6330 Cham, Switzerland

Preface

When I was a child, I read a children's biology book about lizards and other creatures that have the ability to regenerate their limbs and tails; interestingly, such regenerative abilities are almost non-existent when it comes to mammals.

At the same time, as most boys in Japan, I was exposed to many manga, anime, and movies with characters that could regenerate their body parts with special powers. For example, characters from the manga Dragon Ball and the movie Terminator 2 could regenerate their original bodies by aggregating spontaneously after being dismembered by an opponent.

Unfortunately, almost all characters with such regenerative powers were villains.

Then, in high school, I learned in biology class that the dissociated cells of a marine sponge or an amphibian embryo have the capability of spontaneous cell aggregation and tissue regeneration under appropriate conditions. This convinced me that the authors of such manga, anime, and sci-fi movies must have had some background in biology.

After an orthopedic surgery residency, I was a graduate student beginning my research in regenerative medicine when I realized the strong potential of multicellular spheroids in regenerative medicine and tissue engineering; this somehow reminded me of those comic book and movie villains.

While the research history of spheroids dates back to 1907, it seemed to have gone downhill for a while. However, since we invented the Kenzan Method, there has been a renewed interest in "organoid research" in the fields of stem cell research and regenerative medicine.

It has been a decade since the development of the bio-3D printer, Regenova. At the time of my invention, the scaffold-free Kenzan concept was treated with skepticism, treated as heresy, and considered to be outside the golden rules of tissue engineering concepts. However, after 10 years, thanks to many collaborators, we now have several pipelines that are close to clinical applications in blood vessels, peripheral nerves, and so on.

In this book, we describe the history of the Kenzan Method, the development of the bio-3D printer, how researchers have been using it for the development of clinical applications, and an explanation of the targeted organs and diseases including bacterial infection of medical implants. We expect that the readers of this book around the world will develop many new pipelines using our method.

As the cycle of technological innovation is so rapid in the field of regenerative medicine, we are uncertain whether this method will be used as a standard treatment for patients; however, we do hope that this method can help patients suffering from various diseases, especially where surgical intervention and replacement is the last option, as much as possible.

Finally, I wish to acknowledge the support and dedication of collaborators, colleagues, friends, the Japanese government's research foundations, and my family who have made the publication of this book possible.

Saga, Japan   Koichi Nakayama

Contents

Development of a Scaffold-Free 3D Biofabrication System "Kenzan Method"

Koichi Nakayama

Abstract There is a global consensus in tissue engineering that a combination of cells and biomaterials as scaffolds is essential for the artificial creation of three-dimensional (3D) organs. However, our team has combined the existing knowledge of surgical treatment of bone fractures with the classic biological knowledge of cell aggregates to create a completely novel technology for application in regenerative medicine without using biomaterials, named "Kenzan method." Working in collaboration with several manufacturing firms, we have also developed a bio 3D printer. Although many researchers are skeptical of scaffold-free methods, many in vivo studies have shown positive results using our method. In the future, we anticipate that our technology will be able to fabricate transplantable organs in vitro.

Keywords Scaffold free · Biofabrication · Bioprinting · Tissue engineering

1 Introduction

Although organ transplants, in accordance with the law, have been performed in Japan for a few decades, the issue of chronic donor shortage remains unresolved. Studies suggest that the organ shortage crisis is worsening globally, with this phenomenon even being reported in the United States, a major transplantation hub. Some of this shortage is said to be the result of marked improvements in airbags and other safety features, which have reduced automobile accident-related fatalities, the leading source of donor organs [2]. In the face of this limited supply, finding suitable alternatives is of paramount importance. This has meant that methods for

K. Nakayama (✉)
Faculty of Medicine, Saga University, Center for Regenerative Medicine Research,
Saga, Japan
e-mail: Nakayama@nakayama-labs.com

© Springer Nature Switzerland AG 2021
K. Nakayama (ed.), *Kenzan Method for Scaffold-Free Biofabrication*,
https://doi.org/10.1007/978-3-030-58688-1_1

preparing artificial yet functional and transplantable organs in vitro have become a key focus in regenerative medicine research.

In 2001, having completed nearly 4 years of my orthopedic surgery residency at Kyushu University hospital, I entered its clinical graduate school program and was given a research project focused on articular cartilage regeneration. Since the times of Hippocrates, surgeons have known that articular cartilage never heal spontaneously once damaged, and medicine has been striving to find ways to improve these outcomes ever since. In the late 1990s, after the establishment of a chondrocyte culture method, one of the earliest breakthroughs in the field of regenerative medicine was the realization that you can repair articular cartilage by transplanting in vitro cultured cells into an affected joint surface.

Autologous chondrocyte implantation (ACI) was already the mainstream therapy of choice when I entered the regenerative medicine field and was already in clinical use in Europe and the United States, and both medium- and long-term data had been published [1]. Meanwhile, in Japan, a clinical trial was about to begin for evaluating J-TEC's atelocollagen-based ACI model. These methods essentially consist of endoscopically harvesting a patient's chondrocytes from their own joint(s) and placing them in an expansion culture and then retransplanting them into the patient. It was however apparent that this treatment was limited—in a quantitative sense—by the number of healthy cells that can be collected in the first step. In addition, the collected cells need to be repeatedly passaged in monolayer culture, resulting in chondrocyte dedifferentiation and reduced type II collagen production, which is essential to cartilage health. Thus, because of the limited availability of high-quality chondrocytes, ACI can treat only a relatively small area of the defect. However, most patients suffering from articular cartilage injury have larger size of defect.

Having read a paper that bone marrow–derived mesenchymal stem cells (MSCs) could be induced to differentiate into chondrocytes [8], we decided to use MSCs as cell source for cartilage regeneration research. We were starting from scratch, as no one in our orthopedic surgery department at that time had conducted any experiments or research in regenerative medicine, including stem cell culture.

First, to replicate the published protocol, we attempted to induce the differentiation of chondrocytes from bone marrow–derived rabbit MSCs. Briefly, following expansion in monolayer culture, MSCs were isolated using trypsin/EDTA and recovered in 1.5 cc of chondro-induction medium (1.5 ml). Samples were placed in separate microtubes (15 ml each), centrifuged, and then moved to an incubator for cultivation, taking care not to disrupt the cell pellet. The next day, the pellet was floated from the attached inner surface of the tube by finger tapping gently and then cultured for several more days for chondro-differentiation. This culture yielded a single pellet per tube, but the small volume of medium meant that it needed to be replaced every few days, without losing the cell pellet deposited at the bottom of the long, narrow tube. Creating large quantities of these induced chondrocyte pellets required equally high amounts of microtubes and levels of caution when replacing the medium.

At the time, I shared a single incubator with several others, so the culture method of putting several tubes on a tube stand inside the incubator would take up a lot of space and would require coordination with others, which was cumbersome. This method was literally exhausting.

2   Spheroids (Cell Aggregates) and the Mold Method

In the fall of my first year of graduate school, while I was struggling to mass produce these pellets all alone, I happened to see an advertisement for an ultralow attachment 96-well multi-well plate, claiming that it could effectively construct spherical cell aggregates ("multicellular spheroids") from the human liver carcinoma cell line, HepG2. The design capitalized on the discovery that cell aggregates will spontaneously form when an adhesive cell type is forced into high-density suspension culture [9, 10]. The advertisement also noted that the system was not limited to hepatocytes and could readily be applied to prepare cell aggregates of nervous cells ("neurospheres").

I immediately ordered a sample and seeded the plate with MSCs as soon as it arrived. Next, I observed 96 perfectly round cell aggregates in each of the wells. The cells in these spherical aggregates, which I later confirmed, could be successfully differentiated into chondrocytes with normal chondro-induction protocol.

As I gazed at these spheroids, I wondered what would happen if I placed several aggregates into the same one well. I hypothesized that perhaps the extracellular matrix (ECM) exuded by the chondrocytes would behave like an adhesive and make the aggregates fuse together, potentially facilitating the construction of large, three-dimensional (3D) structures without adding biomaterials. I quickly placed approximately four spheroids into a single well of the 96-well plate and cultured them overnight. The very next day brought an unexpected observation. Normally, when MSCs have been induced to differentiate, it takes approximately 10 days for them to transform into chondrocytes, but the spheroids I had put into the well already appeared to be adhering to one another and even fusing less than 24 h after the experiment began. Within the next 2 days of culture, they had completely fused to form a single large colony. Surprised by this behavior, I speculated that if these spheroids could be applied in sufficiently large quantities, scientists could engineer functional 3D structures consisting of cells alone, with no need for a scaffold. In the future, even joint-like structures with curved faces might be possible. At this time, tissue engineering remained the dominant paradigm in regenerative medicine, aimed at synthesizing 3D organs by making cells adhere to polymers or other scaffold material(s) [4]. However, it was difficult to obtain the right kind of material—a high molecular weight, biodegradable polymer that would remain stable long term, without being hydrolyzed in culture medium—since no such product was commercially available at the time.

I seeded each well for spheroid formation with fewer than 1×10^3 cells isolated from the expansion culture at that time. When I tried to prepare giant spheroids by adding 1×10^6 cells to each well, resulting in a 1000-fold more cells in a single well, after a few hours, the medium had turned yellow; and by the next morning, all cells were still floating free in the media. This suggested that most of all cells were dead, possibly due to nutrition shortage, having failed to form aggregates. Next, I went back to the original protocol, seeding 1×10^3 MSCs/well, and when I added all of the resulting spheroids into a single well for further incubation, they bound together

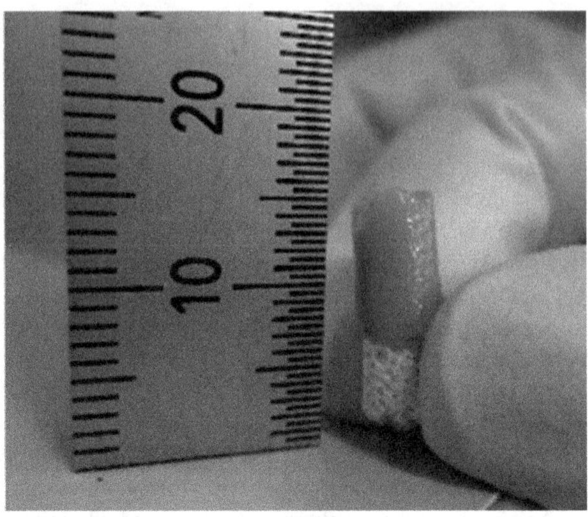

Fig. 1 Cylindrical structure, consisting of only cells, placed on a support platform (white), fabricated using the "mold method"

by the next day. I believed that these results signified the nutritional demands of these cells, which were reduced once they formed these spherical aggregates.

Next, I cut a hole at the bottom of the cryotube, and after sterilizing it, I poured the spheroids into the tube surrounded by a larger culture dish. By the next day, the cells had bound together to form a 3D structure that coated the inner wall of the tube (Fig. 1). Several papers had been published prior to my observations, each claiming to have engineered cylinder-shaped structures consisting only of cells, but each of these methods required a culture period of several months, and their thickness (height) maxed out at around 2 mm. However, with our new "mold method," we could construct structures that were up to 1 cm thick, if the cell conditions were right.

After evaluating several parameters by trial and error, we arrived at a protocol that could reproducibly create stable, cylindrical structures consisting solely of bone marrow–derived MSCs from a rabbit. Then we implanted this construct into the 4.8 mm diameter of small osteo-chondral defect at knee joints of rabbits as autologous implantation and later confirmed smooth articular cartilage regeneration with hyaline-cartilage and subchondral bone [3, 5]. Unlike other osteochondral regeneration systems, our expanded, spheroidized, and still undifferentiated MSCs differentiated correctly—into bone in the bone layer and cartilage in the cartilage layer—once implanted at the joint site. These implants were even able to replicate the characteristic microstructure of articular cartilage tissues.

After rabbit and mini-pig trials with adipose-derived MSCs, we then carried out clinical study in humans from 2016 to 2017. The purpose of that clinical study was to confirm the safety of using the mold-made scaffold-free construct. There were no adverse events observed; however, surgical protocol must require modification for future application.

As articular cartilage is a relatively hypoxic tissue, we believed that this approach—despite poor medium circulation during culture (which is generally unfavorable to cell proliferation)—will still successfully deliver 3D structures consisting of cells alone. However, the prospect of manually preparing mass quantities of spheroids and pouring them into the molds, while sure to work, would be unnecessarily labor intensive. Therefore, I started to wonder if this process could be automated somehow. At this time, a photocuring-based, rapid prototyping system was set up in a joint laboratory within the Kyushu University School of Medicine. Bearing this in mind, I wondered if it was possible to design a device that could "stack" spheroids automatically. I wondered if the application of the principles behind rapid prototyping (namely, additive layering) could be leveraged to produce our larger cell-only products.

Since the Mold method has some limitations, such as it is difficult, in theory, to build a hollow-shaped construct or even mix different types of cell spheroids into a single-shaped construct. By stacking up spheroids one on top of another, like building blocks, we could perhaps construct hollow structures or even further generate organ-like tissues consisting of only cells by stacking various cell types into a predesigned configuration.

I decided to ask for corporate support, as I had no knowledge of mechanical engineering or control system design. I proposed the device concept to famous robotics makers and several manufacturing firms and proposed a technological partnership to facilitate the development of this device. However, the plan had a fatal flaw—my limited research funding as a graduate student—and so I was rejected by almost all the companies I approached.

Finally, after almost 2 years and visits to numerous manufacturing firms, two companies agreed to help me develop my device: a local biosystems manufacturer and clean bench reseller and an arcade game venture business that used bipedal robots. We applied for joint research funding (a JST grant-in-aid) and we were accepted, heralding the actual start of the development phase of this technology. During discussions with our corporate partners, the robot venture business brought an interesting idea to the table, namely, why not manipulate the spheroids by piercing them with a needle? Initially, this proposal was rejected, written off as preposterous and a strange idea put forth by an engineer ignorant of matters of cells and biology.

3 External Fixation

During the kick-off meeting of the funded project, our teams had debated on how to layer the spheroids in practical terms. As complex shapes were difficult to construct using the mold method, the consensus was that a different technique would be necessary. However, we were stumped. Day after day, no specific solutions came forth. Moreover, the collaborator companies did not have a strong passion for realizing our project and did not contribute with any useful ideas. As the project was already accepted for government funding, and as the leader of the project, I was not allowed to give up, soon I realized that I had to do everything alone.

One day, I was attending a postoperative conference in the Department of Orthopedic Surgery, Kyushu university hospital when my gaze landed on a textbook chapter that a medical student had been poring over. The chapter detailed several techniques for treating bone fractures. Normally, a fracture is treated by forcing the displaced fragments back into their original positions ("reduction") and immobilizing them using a cast or similar instrument. Ideally, this procedure should be performed noninvasively whenever possible. If the reduction step is performed suitably, the bone(s) will fuse together after 4–8 weeks of immobilization and the patient can begin rehabilitation. However, when the displacement caused by the fracture is very severe, surgical intervention becomes the only option. In these cases, doctors usually choose *internal fixation*, an approach in which the broken bone(s) are repositioned (more or less) in their anatomically correct location and then fixed using plates, screws, pins, or other tools made of titanium or stainless steel. After a child younger than 10 years old breaks a bone, even if it re-fuses slightly bent or somewhat askew from its original position, it will straighten out completely as they grow and develop, becoming indistinguishable from normal bone in just a few years, thanks to a process known as bone remodeling—even doctors except orthopedic surgeons are sometimes surprised by this fact. Approximately half a year after a bone is treated with internal fixation fuses, doctors typically remove the implants via an operation called hardware removal.

However, in very severe cases there is another category of surgical techniques for treating fractures called external fixation [7] (Fig. 2). This approach involves embedding pins into the bone through the skin, which are then immobilized outside the body. This approach can be extremely effective, depending on the pattern of the break, because it can immobilize the area around the fracture directly and it is not necessary to expose the fractured site. It also involves much smaller incisions into the skin and muscles surrounding the fracture, making it more convenient than conventional plate fixation in suitable fracture cases.

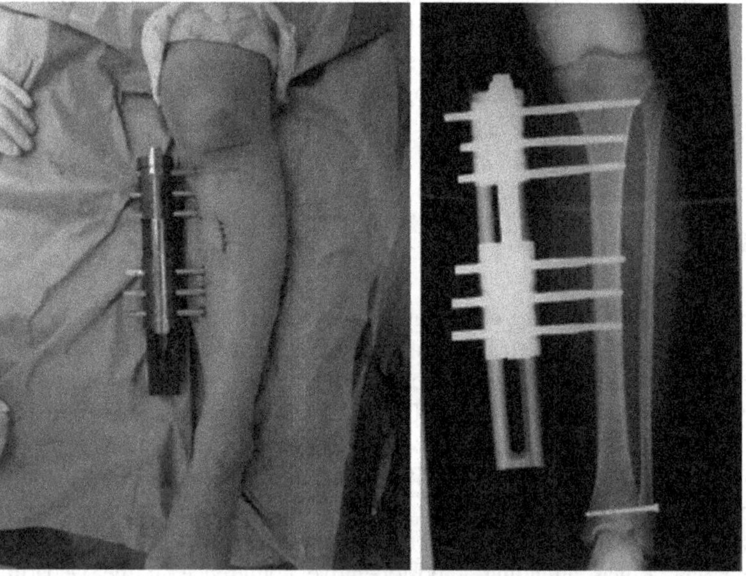

Fig. 2 External fixation (By courtesy of Prof. Nakashima, Orthopaedic surgery, Kyushu university.)

As I pored over the chapter (having appropriated the textbook from the student), I had an idea; why not temporarily immobilize the spheroids in place? The working principle at the heart of surgical therapies for bone fracture is that once the bone fragments are repositioned and fixed—assuming a suitable (i.e., clean, nutrient-rich) tissue microenvironment—the body's self-healing ability will kick in and the patient merely has to wait for the bone to fuse back together. The treatment is over once doctors remove the fixator, and the patient can begin rehabilitation in earnest. I arrived at the tentative theory that our spheroids could similarly bind or fuse together if they were temporarily immobilized with a minimal degree of injury to the spheroids themselves and their constituent cells.

Combining this with the idea floated by the robotics engineer around half a year earlier, my idea was to create a spheroid "skewer," threading them along a thin needle to force them to stay together. Right away, I unraveled some copper electrode wire I had on hand and tried to "thread" a few spheroids on to it. After much struggle and frustration, I successfully made my first skewer (Fig. 3, left). Because of the potential cytotoxicity of the copper ions released from the wire, I then searched around for other candidate materials. The first to draw my attention were toothbrush bristles, which are biologically inert. I purchased several kinds of toothbrushes advertised as "very fine" at a nearby supermarket, removed a single bristle, and attempted to impale a few spheroids on it under a stereomicroscope. Somehow, I succeeded (Fig. 3, right), although it took 2 or 3 h just to place two spheroids on a bristle, because the fiber was too flexible. I put this "skewer" in a 10-cm culture dish filled with medium and left it in the incubator overnight. By the next morning, the dish had turned a vibrant yellow, indicating bacterial contamination. I realized that despite being advertised as having an "antibacterial coating" (among other features), the bristles would need to be sterilized. However, autoclaving was out of the question, since they were overwhelmingly nylon. These challenges meant that they were not the ideal material and I decided to search for other options.

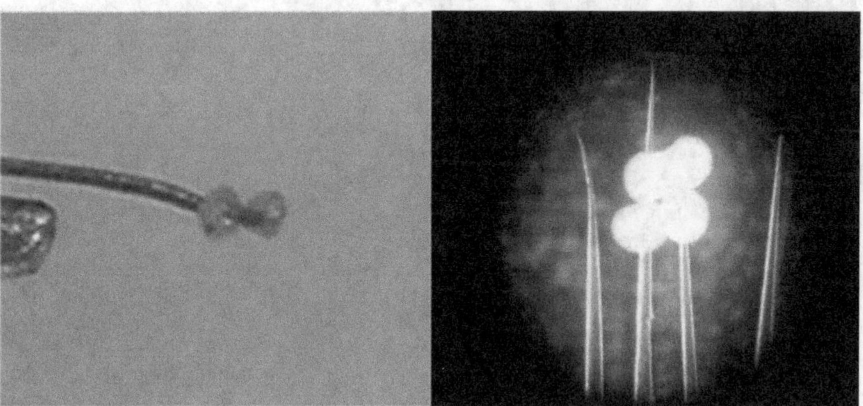

Fig. 3 Aggregates impaled on a wire. (Left) Electrode wire. (Right) Tip of nylon toothbrush bristle

There was no needle thin enough to pierce the tiny ~0.5 mm-diameter spheroids to be found at my local supermarket, DIY center, or arts and craft store. About this time, I was one day randomly leafing through a medical magazine lying in the outpatient waiting area of our orthopedic surgery department when I came across an article written by the president of Okano Kogyo, a small Japanese manufacture company. The article described Nanopass®, an injection needle developed by his firm, advertising them as the world's thinnest needle (now manufactured and sold by Terumo Corporation). This was exactly what I needed, I thought, and so I rushed to look up the firm's phone number online and called them up. Imagine my surprise when President Okano himself came onto the line. Point blank, I asked him for some plain (untreated) needles, and he assented in a matter of seconds. Once the package from Terumo arrived, I sterilized one of the needles and "strung" a few spheroids along it. The next day, I was greeted by the sight of a successful fusion.

It was several more months of struggle before I managed to organize the needles in a 3 × 3 square grid skewer mesh filter membrane at the bottom as remover and impale the spheroids on them under the stereomicroscope (Fig. 4) in a final shape that looked like the "Ikebana" (Japanese flower arrangement) device called "Ken-Zan" (floral pin frog). Now that we could construct 3D assemblies consisting of only cells in pretty much any shape we desired, I realized it might be possible to apply this approach outside the orthopedic surgery field, that is, we could use cells other than chondrocytes and MSCs to answer a different set of questions.

We tested this idea by acquiring some cardiomyocytes, fabricating beating spheroids from them, and skewered them onto the 9 × 9 needles of Kenzan. By the next day, the spheroids—which had been beating independently of one another—had synchronized to the same pulse throughout the structure. This finding transformed the potential of our "Kenzan skewer method" in an instant.

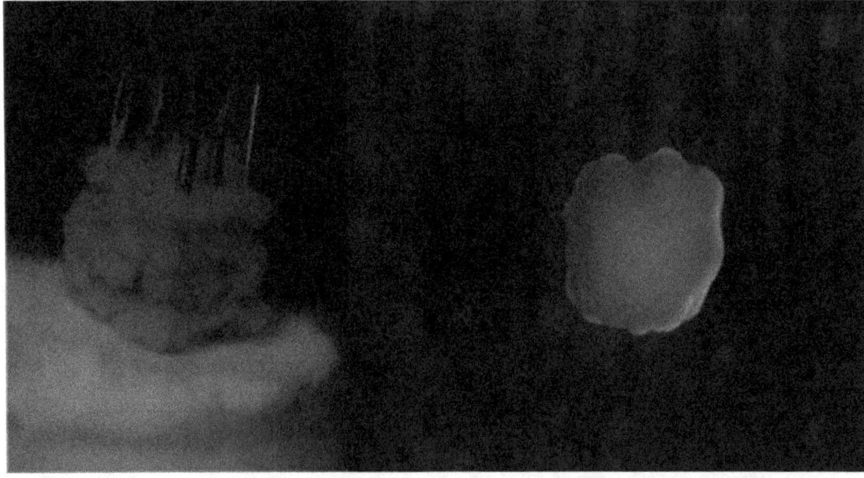

Fig. 4 (Left) Spheroids impaled on hand-made 3 × 3 Kenzan. (Right) 3D cell structure obtained the next day, after removing the Kenzan

4　The Idea of Automating the Capture, Skewering, and Placement of Spheroids

Despite this proof of concept, the task of manually skewering spheroid after spheroid with a diameter of 0.5 mm (or even, in some cases, 0.1 mm) on needles, based on the specific layout desired, would be so time-consuming as to render this application useless in a clinical setting. In fact, we suspected that this extremely long process was the reason for the high rate of bacterial contamination despite performing this step under a stereomicroscope in a clean bench.

I had finally established the basic operating principle of the technique, but we knew it would not be feasible to manually place the spheroids in any scaled setting. I wondered if we could not automate the process and that perhaps our corporate partners may be able to do so; unfortunately, our corporate partners told me that they were very busy and had no time to spare for automation-related R&D.

Time was slipping away. One day, a photo of a microinjection system for in vitro fertilization printed on the cover of a scientific journal caught my eye. After reading a bit more, I learned that it worked by securing the ovum, with a precision in the order of microns, and piercing it with an ultrafine glass needle. This was essentially the same process we were trying to create. We contacted the local company branch of Eppendorf Japan that was selling the microinjection system and asked if we could borrow the microinjection system for a trial period, and they agreed to give it to us for a week. While experimenting with the device—which was operated using a joystick—I noticed an RS-232C port (traditional computer interface similar to an USB interface) on its rear face. Although there was no mention of this interface in the Japanese instruction manual, the original English manual gave a list of RS-232C commands, indicating that it would be quite simple to operate this device at the PC script level (a simple list of commands for computer). When I was an elementary school student, I remember learning programming enthusiastically on an entry model computer that was being demonstrated at a nearby electronics store in the late 1970s. At that time, computers were not as powerful as today's computers and the graphics performance was extremely low. Even though I had been building my own PCs since becoming a doctor, I had absolutely no knowledge of how to program for Windows; the only relevant experience I had was with BASIC and Assembler (both are old computer programming languages) in elementary and junior high schools long before the appearance of Microsoft Windows, or even MS-DOS. As I searched online for a simple programming solution that could give us RS-232C control, I noticed that the "Express" version of the developer software for Microsoft's Visual Basic was now offered free of charge under certain license. That kind of programming development software cost a couple of 100 dollars a few years prior.

Despite some limited features in this software package, it could still accommodate all the operations I was hoping for. Especially thanks to the Internet and programming community, I found many tutorials and sample codes for what I wanted to do. I downloaded the development environments I needed and bought several primers at a bookstore. After several months of trial and error, I could freely control the microinjection system from the computer.

Then I purchased a second-hand Sony's first generation consumer-use HD video camera, which were just starting to catch on at the time. I attached a macro lens on the HD cam and connected it to the PC via FireWire (a high speed PC interface in those days). Using this approach, we were able to image the spheroids in real time on the PC monitor without an objective microscope, at a much higher image quality than I had expected. At that time, I hired a post-doc who had majored in mechanical engineer. He supported with the hardware for developing my prototype and attached a Sony PlayStation controller to the system for fine manual control of the head of the microinjector. I set up the video camera on a camera platform stand (often used for specimen photography) and attached an infrared (IR) emitter near the camera, which was completely programmable via PC. Instead of using the original IR remote controller for the video camera, I tried to handle image capture and scaling via a program code. However, since the remote controller had no function to adjust the focus, you had to manually rotate the focus ring on the lens of the camera body. I thought it would be difficult to adjust the focus via PC unless I disassembled the camera body and modified it electrically. That was technically impossible for the two of us in those days. Then I came up with a makeshift solution. I purchased a cheap IR-controlled toy car, disassembled the body, then attached the IR-controlled motor-tire system to the bottom of the camera stand, and connected to the focusing ring through a rubber belt. Finally, we were able to adjust the image focus via the IR emitter. Later the engineer replaced the toy car body motor with the stepping motor directly connected to the PC. Finally, we were able to pick up spheroids using the gaming controller and by watching through the PC monitor. This presented an escape from the mind-numbing, sensitive work of directly grasping the spheroids with conventional ultrafine tweezers or pipettes to skewer spheroids onto a needle in the Kenzan. Systems capable of displaying a field of view from a biological microscope onto a computer screen were available, but they were prohibitively expensive. Our prototype was much cheaper but could do more.

Nonetheless, all that had really changed was the means of manipulation, from tweezers to mouse; the task was still as labor-intensive as before, as I still needed to manually, visually, and individually arrange the spheroids. While I was looking through some informational websites devoted to programming, I came upon a special feature on image processing and recognition. In addition, I came across the Japanese edition of "Make magazine (O'Reilly Japan)," first published in those days. I was shocked to learn that OpenCV—an open source project loaded with facial recognition, objective tracking, and other futuristic technologies, operable even when using an inexpensive USB webcam—was now available free of charge. I downloaded it right away, and as I experimented with the sample program, I became convinced that spheroid recognition could be automated and the whole fabrication process could be automated probably by myself.

Although it took me approximately a year of testing and tweaking this simple microscope–PC system, eventually the system could automatically detect the center coordination and radius of spheroids, the positions of the tips of the needles in the Kenzan, and other measurements using a combination of relatively simple image recognition algorithms. Nearly all the biology and medical books had disappeared

Fig. 5 Original bio 3D printer (self-built prototype)

from around my desk, which was now buried under mountains of programming- and PC-related books. Visitors to my office would often make jokes, wondering aloud whether I was really a doctor. Between my outpatient duties and conferences in the orthopedic surgery department, I took every moment I could to continue to program and test my system. Nearly 2 years after starting the project, the system had improved enough for a simple demonstration, automatically recognizing, capturing, and skewering spheroids onto desired needles in the Kenzan without using the gaming controller.

This prototype looked like a poorly cobbled together pile of junk but was more than up to the task of proving the concept behind our innovation (Fig. 5). Just as I was putting this working demonstration into place, I was visited by a few technicians from a company I had never heard of, sent to me by Dr. Hajime Ohgushi at AIST (a Japanese government–funded research institute), the Shibuya Corporation. This company lacks name recognition because none of its business areas serve the general consumer, but it is well known by those who require high levels of technical expertise; their products span beer and juice filling lines, semiconductor manufacturing equipment, dialysis machines, and original equipment manufacturer (OEM) biomedical isolators.

They were visiting on an unrelated matter but showed keen interest in the prototype and the Kenzan method, which I demonstrated during a tour of our lab. They were very knowledgeable about the current state of affairs in regenerative medicine—especially the conventional wisdom in tissue engineering circles that it is impossible to create 3D organs without the use of a scaffold material—and thus, they were deeply intrigued by our unusual technique of constructing 3D cell structures without biomaterial. It was at this point that they offered to improve and refine my prototype, clearly the work of an amateur, using the Shibuya Corporation's pro-

Fig. 6 Regenenova®: the completed bio 3D printer, developed in partnership with Shibuya Corporation (sold by Cyfuse Biomedical)

prietary technology. Accordingly, I took the opportunity to visit their headquarters in Kanazawa. After nearly half a year of meetings dedicated to ironing out the technical specifications, we managed to create a finished product [6] (Fig. 6). While optimizing and updating the machine and protocol, we tried to create various structures using different types of cells, and we were surprised at the precision, performance, and results the system was able to achieve, all of which exceeded our expectations.

After successfully automating the Kenzan method, there were many things about the 3D-printed results and outcomes that surprised us. With the first machine codeveloped by Shibuya Industry, two types of spheroids could be arranged in a free 3D position. However, one day, when the machine was used to print with only one kind of heterogeneous spheroid mixed with several cells, we noticed that the cells self-organized in the Kenzan after a few days of culture. We assumed that these printed cell–cell interactions and culture environments determine the migration of the Kenzan printed cells. For example, when we printed endothelial cell (EC) and fibroblast-mixed spheroids into a tubular construct, histology showed a capillary-like structure and the inner luminal layer was covered by ECs, indicating that the cells had migrated from each spheroid. Another example printed with mixed hepatocyte, EC, and fibroblast showed spontaneous formation of liver-like structure such as sinusoids-like structure and Glisson's sheath-like structure. This phenomenon of cell self-assembly in the construct made by the Kenzan method convinced us that for future organ regeneration, it is not necessary to accurately place individual cells

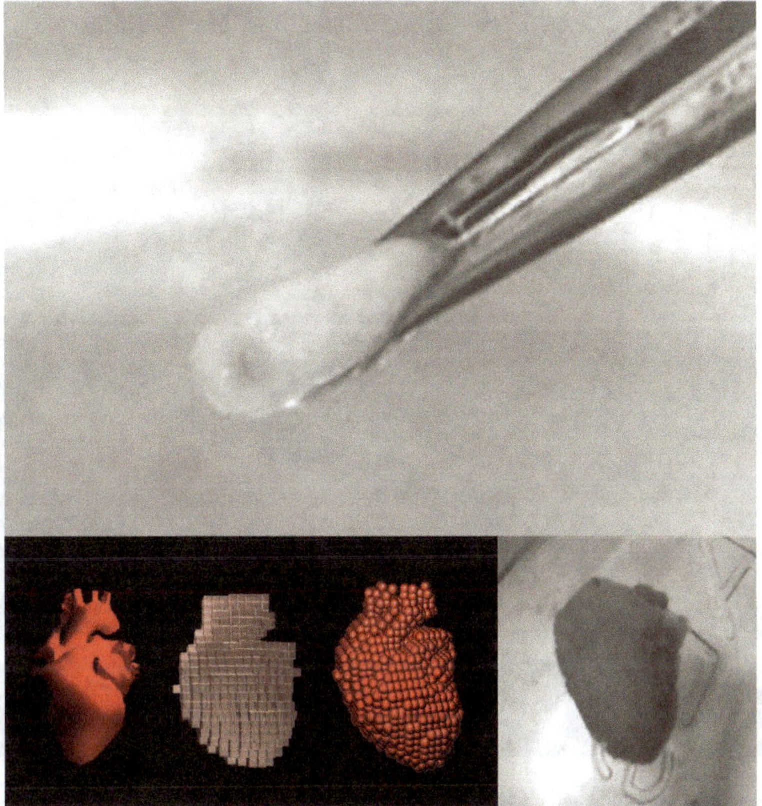

Fig. 7 Cell assemblies fabricated using Regenenova®. (Top) Hollow, vessel-like structure (vascular endothelial cells and fibroblasts). (Bottom row) Heart-like structure, fabricated by layering cells based on 3D scan data from a real heart. Note: Since the cells are only fibroblasts, the structure does not 'beat' like a cardiomyocyte complex would

one by one in three-dimensional space according to its anatomical structure, at least by using our method (Fig. 7).

Another phenomenon that particularly surprised us was the connecting of the printed structures to each other (Fig. 8). From the time I conceived of this method, I thought it would be possible to fabricate tubular organs with a layered structure. In fact, we had been successful in printing tube-like structures that were approximately 5 mm long and 5 mm in diameter, but we thought it would be difficult to make longer cell tubes with a Kenzan needle array, which was approximately 5 mm in height at the time. This was because each needle of the Kenzan that we used was a stainless steel needle of 0.1 mm diameter, and if it is too long, the needle tip will deflect and become unstable under its own weight, making it extremely difficult to pierce the spheroid with the machine.

However, one day, one of the collaborating researchers happened to find that two independent cell tubes connected nicely when they were put together and attached

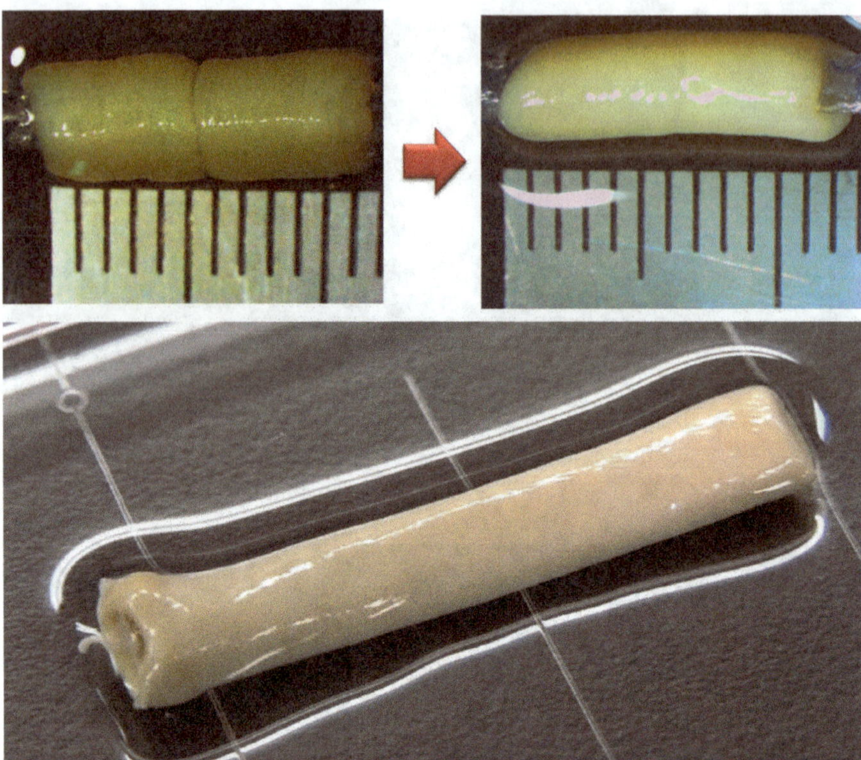

Fig. 8 (Top) Connection and elongation of cell tubes (bottom) cell tube of 7 cm length and 6 mm inner diameter made by the connection method

after a few days of the print. The printed structures connected and smoothly fused with each other within a few days of being printed. I believe this is because there is only little secretion of ECM in the structure immediately after printing. Notably, with increased amounts of ECM in the structure, the smooth connection did not occur. These results convinced us that, given a sufficient number of cells, we could theoretically make any number of long cell tubes.

In general, cellular structures with a thickness of more than 200 µm are considered to be necrotic owing to the absence of capillaries with blood flow. In fact, even the cellular constructs printed by our method, especially the constructs made from cardiomyocytes or hepatocytes, easily necrotize when the interior is densely printed. However, because our method can easily place spheroids in a 3D space, when dealing with such nutrient-demanding cells, we have found that the internal cells can survive even if the construct thickness is larger than 1 cm by designing them to have small internal flow channels; this strategy can even maintain metabolic functions when the construct is made from hepatocytes.

Although I struggled to find early support for this project, we now have the support of a growing number of professionals and several joint research projects between my lab and others, as well as corporate collaborations, underscoring the importance of industry-academia-government cooperation.

Looking forward, my day-to-day research activities will continue to focus on using our bio 3D printer in clinical settings, as a new, patient-friendly, medical technology facilitating new therapeutic approaches. This technology is superior to the preexisting ones in terms of cost and safety. With this in place, it will not be long before we can realize our ultimate goal of creating synthetic organs for autologous transplantation, combating the problems of organ rejection and shortages across the world.

Acknowledgments We would like to thank Editage (www.editage.com) for English language editing.

References

1. Brittberg M, Lindahl A, Nilsson A et al (1994) Treatment of deep cartilage defects in the knee with autologous chondrocyte transplantation. N Engl J Med 331:889–895. https://doi.org/10.1056/NEJM199410063311401
2. International Summit on Transplant Tourism and Organ Trafficking (2008) The declaration of Istanbul on organ trafficking and transplant tourism. Clin J Am Soc Nephrol 3:1227–1231. https://doi.org/10.2215/CJN.03320708
3. Ishihara K, Nakayama K, Akieda S et al (2014) Simultaneous regeneration of full-thickness cartilage and subchondral bone defects in vivo using a three-dimensional scaffold-free autologous construct derived from high-density bone marrow-derived mesenchymal stem cells. J Orthop Surg Res 9:98. https://doi.org/10.1186/s13018-014-0098-z
4. Langer R, Vacanti JP (1993) Tissue engineering. Science 260:920–926. https://doi.org/10.1126/science.8493529
5. Murata D, Tokunaga S, Tamura T et al (2015) A preliminary study of osteochondral regeneration using a scaffold-free three-dimensional construct of porcine adipose tissue-derived mesenchymal stem cells. J Orthop Surg Res 10:35. https://doi.org/10.1186/s13018-015-0173-0
6. Nakayama K (2013) In vitro biofabrication of tissues and organs. Biofabrication:1–21. https://doi.org/10.1016/B978-1-4557-2852-7.00001-9
7. Paul GW (2003) The history of external fixation. Clin Podiatr Med Surg 20:1–8, v. https://doi.org/10.1016/s0891-8422(02)00050-2
8. Pittenger MF, Mackay AM, Beck SC et al (1999) Multilineage potential of adult human mesenchymal stem cells. Science 284:143–147. https://doi.org/10.1126/science.284.5411.143
9. Townes PL, Holtfreter J (1955) Directed movements and selective adhesion of embryonic amphibian cells. J Exp Zool 128:53–120. https://doi.org/10.1002/jez.1401280105
10. Wilson H (1907) On some phenomena of coalescence and regeneration in sponges. J Exp Zool 5:245–258

Position of the *Kenzan* Method in the Space-Time of Tissue Engineering

Nicanor I. Moldovan

Abstract The Kenzan method is a bioprinting method which has already generated several of the most relevant in vivo applications of tissue engineering and holds promise for further substantial development. However, it is less appreciated that Kenzan's success derives from being one of the most truly "engineering" methods among the robotic biofabrication technologies. This is not due only to the exquisite sophistication and precision of the two instruments operating on this approach (*Regenova* and *S-PIKE*) but also to the very principle it implements in practice: the integration of the biological, modular, and semistochastic space–time of the cell spheroids within the Cartesian space created by the rectangular microneedles array. For this reason, after an overview of the Kenzan method in the larger context of bioprinting, here we will comparatively discuss these two frameworks, and then we will illustrate their synthesis with examples generated by virtual computer simulations.

Keywords Kenzan · Scaffold-free · Spheroids · Self-assembling · Bioprinting · Tissue engineering

1 The Kenzan Method and Bioprinting

The principles of the Kenzan bioassembling method as well as its supporting technology and applications have been previously discussed in detail [1–3]. To summarize, this tissue engineering approach, first implemented in the *Regenova* instrument,

N. I. Moldovan (✉)
Indiana Institute for Medical Research, Indianapolis, IN, USA

'Richard L. Roudebush' VA Medical Center, Indianapolis, IN, USA

IUPUI, Indianapolis, IN, USA
e-mail: nimoldov@iupui.edu

© Springer Nature Switzerland AG 2021

K. Nakayama (ed.), *Kenzan Method for Scaffold-Free Biofabrication*,
https://doi.org/10.1007/978-3-030-58688-1_2

is based on the use of metal microneedles called "Kenzans" [4] for robotically lacing cell spheroids in 3D patterns. Using this method, a variety of tissue-like constructs were created in vitro and also tested by successful in vivo implantation: vascular grafts [5, 6], cardiac patches [7, 8], cardiac [9], tracheal [10], urethral [11] and esophageal [12] tubes, liver mini-tissues [13], and others.

The condition for the cell spheroids to fuse in the Kenzan needle array is placing and keeping them long enough in contact, which requires that the spheroid dimensions are constrained to a uniform size. To give more freedom to the users in spheroids selection, the "next-generation" instrument based on the Kenzan principle (*S-PIKE*, also commercialized by Cyfuse Biomedicak, KK), uses individual, mobile microneedles that penetrate the spheroids like toothpicks, and brings them in contact by immobilizing the microneedles on a soft supporting material in a computer-determined position.

Of note, the *Regenova* instrument has been prudently called a "Bio-3D Printer" rather than a "bioprinter" [9], signaling that it has unique features which makes it distinct from other bioassembling technologies. In this case, one may wonder if Kenzan bioassembling is a "true" bioprinting method. The answer is a straight "yes," for at least two reasons: the first is that both instruments made so far to implement this method fulfill the criteria of "additive manufacturing," namely, the generation of 3D structures using computer-assisted robotic devices operated in a layer-by-layer organization pattern (more horizontal in Regenova, and more vertical in S-PIKE). Combined with the notion of live cell spheroids as "bioink" (e.g., as in [14]), the Kenzan method should indeed be considered a bona fide "bioprinting" technology [15].

This outstanding technological performance emerged at the confluence of two Japanese traditions: one, to employ needles for fixing a structure in place (as in the Ikebana flower art [16]); and second, that of actual additive manufacturing, as the first patent on 3D printing was actually filled in 1980 by the Japanese engineer Dr. Hideo Kodama from Nagoya Municipal Industrial Research Institute, who described first the concept of "rapid prototyping" methods of plastic models with UV laser photo-hardening of a thermoset polymer, the active area being defined either by a mask pattern or by a scanning fiber transmitter [17, 18].

Since many scientific breakthroughs have multiple tributaries, 3D bioprinting is not different: apparently, when the field was seeking a means to assemble cells spheroids in 3D structures, it might have been considered the use of metallic needles as support, but these were dismissed as too difficult to put in practice, compared to the fugitive hydrogels (which with slight variations became the core of both "3D Bioprinting Solutions" [19] and of "Organovo" [20] companies' technologies). Yet unsurprisingly, the exquisite Japanese technical acumen and dedication to detail made possible the development of powerful robots capable to operate with submillimeter cellular clusters according to the Kenzan method, and thus redefined the bioprinting landscape. Naturally, this method has its own limitations discussed elsewhere [1, 2, 21], in part addressed by other emerging technological alternatives [22].

2 The Kenzan Method and the "Scaffold' Debate

The field of tissue engineering is broadly divided into two complementary and often overlapping approaches, depending on the involvement or not in the constructs of a biomaterial scaffold for maintaining the cells [21]. If the 'scaffold" is considered the supporting hydrogel used as "bioink," then the Kenzan method is a "scaffold-free" technology [5]. The Kenzan microneedles *stricto sensu* do represent a scaffold – although a temporary one – and thus the method might not be perceived as being "purely" scaffold-free. However, considering that the use of hydrogel rods as temporary supports for cell spheroids assembling did not disqualify other bioprinting methods for being defined as "scaffold-free" [23, 24], the same argument should conceptually be applied to the Kenzan method as well.

However, this still depends on what is understood by "scaffold," because as elsewhere in biofabrication, the terminology is in a fluid state, which triggered several attempts to make it more uniform [25, 26]. Obviously, even if the assembling method is scaffold-free, the cells will still need an extracellular matrix (usually produced by themselves). In the extreme, the notion of "scaffold" was limited to *polymeric fibrillar materials*, and then, in an even more debatable acceptance, the hydrogel-based bioprinting itself was called "scaffold-free" [27].

Alternatively, the Kenzan method could be conceived as an advanced form of "molding." A variant of this is when the cellular confinement is provided not by a solid support but through a physical field of energy concentrated in a specific location, and with a predetermined 3D shape. An example is the acoustic field employed to manipulate not only single cells [28] or cell clusters [29], but even large-scale cellular constructs [30]. Another case is the magnetic force–based positioning of cells in 3D assemblies. This method has attained a high degree of precision [31, 32] and sophistication [33, 34], with remarkable applications such as the remote bioassembling in space [35], performed on the "Organ.Aut" automatic bioprinter, recently placed on the International Space Station [36]. To the same category belongs the simpler method of magnetic force–assisted assembling of single cells [37] or cell clusters [38] labeled with magnetic particles, a technology also commercialized as "bioprinting" [39] (although this name could be confusing for a nonspecialist audience).

The building blocks for modular scaffold-free tissue engineering do not need to be spheroids [40] but can also be "tissue strands" [41], toroids [42], or honeycombs [43] (designed to generate by stacking perusable channels in the constructs) that have been assembled in meaningful structures, such as vascular tubes [44] and tumor models [45].

Altogether, the scaffold-free methods have their own advantages but also disadvantages compared to hydrogel-based bioprinting [2], which can be compensated by a "hybrid" approach [3, 46]. In this regard, the combination of cell clusters and "sacrificial" hydrogel supports, as practiced in some versions of bioprinting, could be better considered "hybrid" rather than truly "scaffold-free" bioprinting [23, 24].

3   The Kenzan Method and the Biological Space–Time

Molding-based tissue engineering in general, and current bioprinting in special, operate "outside-in" regarding the construct: namely, they seek to fill a predetermined shape (often geometrical, or derived from a medical 3D image) with a biomaterial. If the main goal of the activity is to *organize the cells and not their extracellular support*, this workflow reverses the order of priorities with biology (although it may seem logical, considering the derivation of bioink-based bioprinting from "additive manufacturing" [47]).

However, while the physical "Newtonian" space is external to – and independent from – the object, the biological space is *internal* to the organism, and continuously generated by its "inside-out" growth. However, the assembling on the Kenzan microneedles of cell spheroids generates an interesting situation, where the two modes of space organization, biological and geometric, must structurally coexist.

Although seemingly less organized than some inorganic materials, the biological objects nevertheless have a complex order, determined by the logic of the organism's development [48].

A property of the biological structures is their fundamental randomness [49]. Of note, the noise and/or errors interfering with the intentional order in artificial constructs are different from biological randomness, which is an essential, constitutive property of living matter making possible many of its emergent properties [50]. The internal ordering of biological structures does not lead to the monotonous, translational symmetries encountered in crystals, but has another sort of regularity, based on the progressive serial addition by growth of subunits ("modularity"). Moreover, the self-similar *linear* growth in 2D, as for example the xylem tubes in leaves and blood vessels in the retina, or in 3D such as the tree branches or the airways in the lungs, can be measured by "fractal dimensions" [51]. Determined by the branching patterns driven by their function (in the mentioned cases, the efficiency of fluid flow coupled with optimal distribution of transported molecules). However, the fractal dimension of a structure embedded in biological space should not be confounded with the *dimensions* of this space. In fact, the framework of physical reality is four-dimensional (4D), the 3D space being intimately coupled with time as the space–time continuum. At the common scale (including that of inert materials), space and time can be treated as separate dimensions. This epistemic operation allows them to be addressed independently, generating the dynamic description of natural phenomena as separately dependent on space and on time, thus generating the "deterministic" description of reality. However, in biology this separation between space and time is not possible, since the biological objects are in long term the products of evolution, and in short term of self-assembling, thus essentially *spatio-temporal*, that is "historical" [48].

Of note, the notion of four-dimensionality (4D) is increasingly used in additive manufacturing to describe programmable materials which change their shape after assembling [52]. This concept has been transferred to bioprinting [53], although the similarity is rather superficial, since the post-printing maturation of the constructs to which it was applied affects more profoundly the biological constructs than the inert ones.

In order to observe these distinctions yet maintain the specifics, the tissue engineering methods of biofabrication, bioassembling and bioprinting (including the Kenzan method), could be considered as "3.5D" techniques. This would indicate a more intimate relationship of temporality with spatial distribution than in the "uncoupled" 3D space +1D time = 4D case of the inanimate dynamical systems. There is a precedent to this type of labeling, when a *fractional* degree of spatial freedom of the cells migration within hydrogels [54], or in a micro-engineered environment (pillars, posts) [55], has been coined as being "2.5D."

Positional orientation in biological space, and consequently in that of tissue engineering, is another subtle issue to deal with from outside of the organism (or living construct), since it is actually determined from the inside of the tissues, as indicated before. This problem of orientation in biological space recently became critical in the attempts to chart at single cell resolution all the cells in the human body, the so-called Human Cell Atlas [56]. One of the most interesting solutions was to consider a topological approach based on the (micro-)circulatory system, with "mile 0" placed in the center of the heart and measuring normalized distances along the blood vessels [57].

Additive manufacturing builds the objects layer by layer or by volume-based aggregation methods [58]. In either case, the spatial coordinates of the components, or of the boundaries in the second case, need to be precisely known to determine a bioprinting instrument's "resolution"). With one exception (BioBot Basic from the company Advanced Solutions for Life Sciences that performs the assembling on a rotating platform, for which the control code is written in polar coordinates [59]), the large majority of bioprinters' software for generation of the printhead's path are conceived in Cartesian (orthogonal) x, y, z coordinates. While this system is very useful for hydrogels and easily implemented in practice, it is almost useless when working with cell aggregates, including spheroids, as the cells tend to relocate from the initial placement to their "comfort zones" inside the construct.

Moreover, from the previous discussion, it might be implied that biological space is "*disorganized*" and thus continuous for any practical reasons. However, biological space also has features which make us conceive it as discrete. For example, there are repetitive subunits employed in biosynthesis over and over again: at the most fundamental level are the (bio)chemical groups processed as identical units, such as the amino acids, nucleotides, and many others. At higher level are the genes, chromosomes, ribosomes, organelles, etc. For sure, the cell itself is the most striking manifestation of the basically discrete and *modular* organization of living matter, making the biological space essentially discontinuous. Still, this space is not fully discrete either, because at any level of organization the subjacent structures generate an embedment that smooths out the discontinuities and fills the apparent "gaps' in between the units.

In the Kenzan method, this "discretization" of the biological space in the cell spheroid-derived constructs becomes *intentional,* as the implementation of a "modular assembling" principle, used in other branches of tissue engineering as well (e.g., in [60]). The use of preformed cell spheroids (up to circa 50,000 cells, depending on cell size) was suggested because the manipulation of single cells to generate constructs by "scaffold-free" bioprinting – although in principle possible – would take a too long time [40]. To speed up the assembling process, was also proposed the employment of repetitive preformed structures, for example those obtained by molding in toroidal or honeycomb shapes [61], assembled again using robotic technologies [62, 63]. Altogether, it should be recognized that the Kenzan method's solution of cell spheroids as building blocks perfectly aligns within the general paradigm of the modular tissue engineering.

4   The Kenzan Method in Virtual Space

As a technological undertaking, tissue engineering relies on the general engineering principles, among which the first is the rational design for function-oriented optimization. For this reason, tissue engineering, through its supporting technologies, heavily depends on computer assisted design (CAD) and computer-assisted modeling (CAM). In fact, bioprinting was from the beginning conceived as the transfer of a tissue model from "virtual reality" into a concrete one, the two spaces being in a biunivocal correspondence [47]. The Kenzan method does not make an exception: as a robotic bioassembling method [4], it uses its own dedicated design-assisting computer program, called *Bio3D* [5], which is admittedly simple, yet sufficient for its intention to predetermine the spheroids' positions in the construct.

The prediction and optimization of the properties of cell spheroids–derived constructs has long benefited from dynamic computer simulations (reviewed in [64]). For example, spheroids formation [65] and their aggregation in tubular structures [66] or in cylindroids [67] were simulated using stochastic Monte Carlo approaches. The spheroids stability [68] and binary fusion [69], have been also addressed using an agent-based modeling of the "Cellular Potts" (CP) class [70], on the multiscale modeling platform CompuCell3D (CC3D) [71].

Computer simulation can be applied specifically to the Kenzan method as well. For example, using a similar CP agent–based approach in 2D (as an approximation of the 3D situation), we simulated the impact of available oxygen on spheroids fusion, and on the formation and duration to spontaneous occlusion of simple hexagonal ring-like constructs [72].

CC3D simulation of Kenzan bioprinting can be profitably performed by taking explicitly into consideration the needles in interaction with the spheroids, as well as their initial distribution and cellular composition. Below we will illustrate these considerations with several unpublished examples. In one of them, the arrangement is rectangular, as in the current Kenzan implementation (Fig. 1), with the spheroids of either two separate cell compositions (Fig. 1a) or as a mixture thereof (Fig. 1b). In the latter case, the simulation shows a rapid "cell sorting" process with the more adhesive ones being attracted among themselves and thus moving away from surface to the depth of the construct. Consequently, the superficial layer of cells rapidly becomes populated exclusively with the lower-adhesiveness type, although inside of the construct the cells remains mixed for longer time (Fig. 1b).

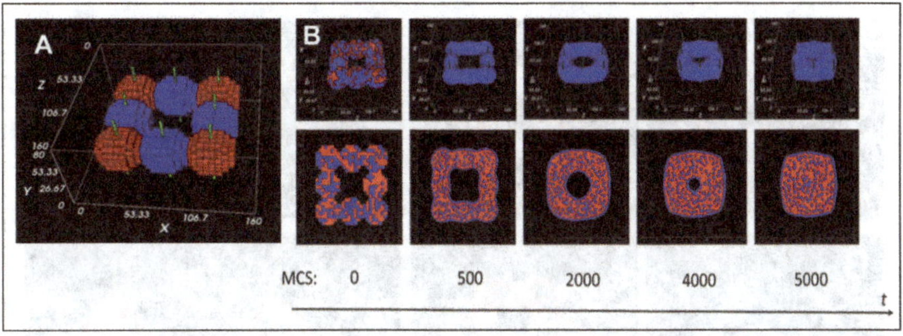

Fig. 1 Simulation of a rectangular array of Kenzan-impaled spheroids made from two different cell types (red = stronger adhesiveness; blue = lower adhesiveness). (**A**). The initial distribution or an array of alternate cellular composition. (**B**). Snapshots from the simulation of an array of mixed-cell spheroids, as seen from the surface (upper row) and from the section at middle height (lower row). Note the fast emergence on the surface of a single-type cell layer (of lower affinity, blue) and the latency of uniform mixing inside of the construct. In this example, the simulation ended at 5000 Monte Carlo Steps (MCS)

This result is interesting, as it has been considered that for explaining the attainment of a state of "minimum energy" equilibrium (and for here, the spherical shape of spheroids and of their resulting fusions), the system would behave like a fluid droplet with superficial tension, i.e., with stronger interactions on the surface than inside [73]. Instead, our modeling suggests in these bicellular constructs a rather "selective depletion" mechanism favoring the more adhesive ones to sink, while the lower-adhesivity cells to be left behind floating on the surface. Given the importance of interfaces in the functioning of biological structures, including in tissue self-organization [74], this observation could be exploited for the optimization of cell distribution in tissue engineering constructs [75]. However, this tendency can be countered by other biological factors, such as the *active relocation* of the cells with lower adhesiveness towards the construct's center, for example by chemotaxis. An example is the formation of mixed endothelial progenitor cells-containing spheroids, which could be attracted at their core by a VEGF gradient, likely produced by the hypoxic conditions found there [76].

Although this might be important for heuristic and validation purposes, the true benefit of virtual simulation is less in reproducing known facts, but more in exploring novel situations not yet tested experimentally. One such situation is a *hexagonal* distribution of the Kenzan microneedles (Figs. 2 and 3), now made possible by the new S-PIKE instrument. The spheroids fusion's evolution can be thus monitored in the virtual space from the outside of the construct (Fig. 2, left column) or in two perpendicular cross sections (Fig. 2 center and right columns). Of note, this simulation highlights the apparent thinning of the spheroids in cross-section as the cells occupy more and more of the inter-needle space, while the construct's inner hole vanishes (compare Figs. 2A-D, right). Moreover, one could observe the tendency of the cells to attach to the needle, leading to an ellipsoidal cross-section of the attached portion (Fig. 2B, C, right), its progressive eccentricity, as well as an unexpected

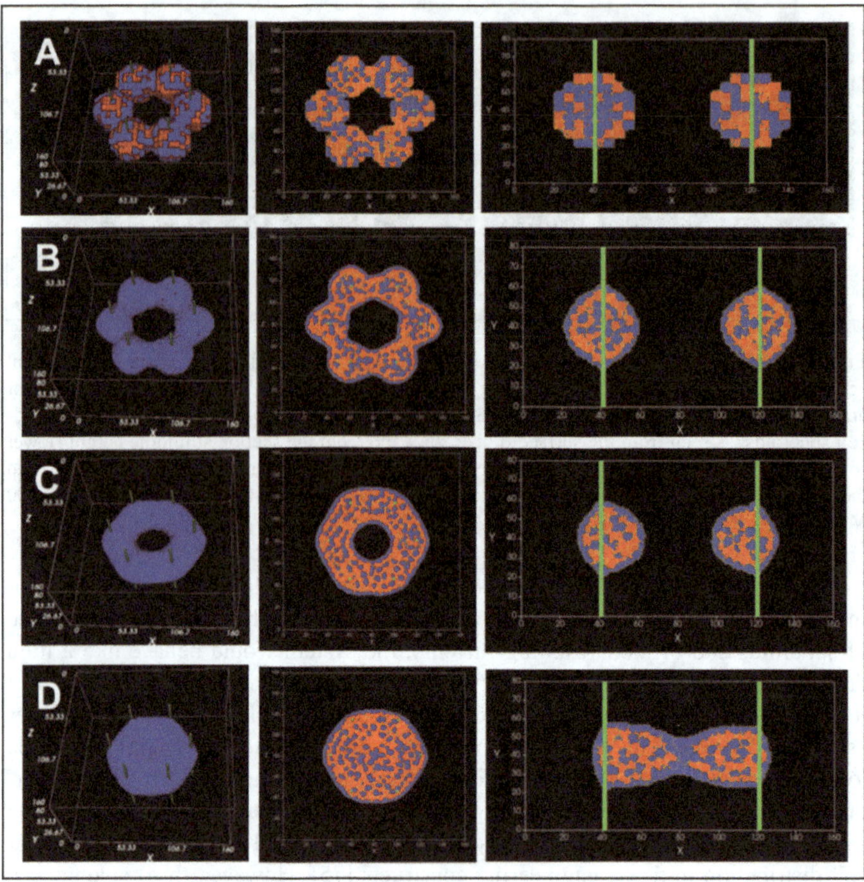

Fig. 2 Evolution of a hexagonal ring-shaped construct as seen in three visualizations (left, from surface; center, horizontal section at mid-height; right, vertical cross-section). Color code and the cells' time-distribution follows those in Fig. 1. Snapshots from 0 (**A**), 100 (**B**), 1000 (**C**) and 2000 (**D**) MCS

latency of cell heterogeneity, that is, the presence of less strongly interacting "blue" cells in the depth of cross-section within the construct (Fig. 2D, right).

An important consideration in these simulations, as in the physical reality, is how the initial conditions impact on their course and outcome. For this reason, we compared two initial spheroid compositions: cells of different adhesivity placed in different spheroids vs. the case when they are randomly distributed in a 1:1 proportion in a mixed composition (Fig. 3). If the fusion process was driven solely by energy minimization, the cell distribution and the construct shape should be the same after enough elapsed time, which is actually seen by comparing cases A and B in Fig. 3.

When the strength of intercellular attachment varies, the final cell distribution will also be different, as shown in Fig. 3C. This corresponds to the case when vascular tubes were prepared by the Kenzan method from spheroids

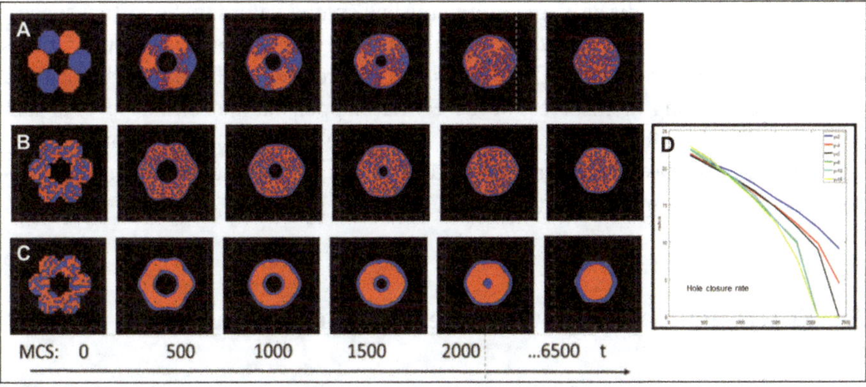

Fig. 3 Comparison of the fusion outcomes of hexagonal structures in three circumstances. (**A**) Fusion of spheroids with different cell compositions, seen from the mid-level section. (**B**) Mixed spheroids of the two cell types. Note the similar cell distribution at the run's end with (**A**). (**C**) Mixed spheroids with larger difference in self-adhesivity: in this case, the "cell sorting," that is, separation between surface and core, is rapid and remains constant. (**D**) Time course of ring's hole closure as dependent on self-adhesivity of strongly interacting (red) cells, increasing from left (yellow line) to right (blue line). Unexpectedly, the strength of this interaction is *directly* related to the hole's latency time (more clearly visible at 1000 and 1500 MSC in B-C). Cells' color code as in Fig. 1

containing stronger-attaching fibroblasts and endothelial cells. These grafts were implanted in animals, and at explantation a continuous intimal endothelial layer was found [5]. This layer could have well originated through the "selective exclusion" mechanism described here, even if these cells were initially intermixed in the medial layers.

In the simulation, we next varied the affinity among the stronger-interacting cells (the lighter-colored cells in Fig. 3B and C), all the other parameters remaining the same. This model offered us the opportunity to compare the rates of closure of the central hole, measured through its diameter, as dependent of the intercellular adhesiveness. Surprisingly, this was *slower* with the increase of homologous affinity among the stronger attaching cells (Fig. 3D), showing that sometimes our intuition could be tricked in more complex situations like this.

For tissue engineering experiments with the Kenzan method, the closure of holes is an important process: it is very desirable when the hole left behind by the microneedles needs to – and largely does – "heal." Alternatively, this closure might be *undesirable* when the lumen of a ring or tube should stay open, such as for the vascular, intestinal or bronchial tubes made from stronger-attaching smooth muscle cells and/or fibroblasts, mixed with lower-affinity endothelial and/or epithelial cells [77]. The simulation suggests that *reducing* the adhesivity between the stronger-attaching cells could delay the hole (or lumen) closure.

Importantly, the experimental validation requires the calibration of these simulation platforms. In CC3D, this has been done for single (tumor) spheroids [68] and for fusions of spheroids prepared from mesenchymal stromal cells [69]. While a *spatial* calibration by directly assigning physical spatial values to the virtual space's voxels is straightforward, the *temporal* calibration is much more difficult. This is due to the fact that the Monte Carlo Steps (MCS) are not ontologically identical (or even similar) among themselves, because what happens at the cellular level is substantially different from one "step" i.e., one cycle of simulation, to another. For more details, we leave the interested readers to consult the references on the problem, and on the attempted solutions (e.g., [64, 78]).

5 Emerging Applications of the Kenzan Method

Understandingly, the most frequently considered utilization of the Kenzan method is for tissue engineering, and ultimately for generation of constructs for either in vitro testing or for in vivo implantation. However, the exploration areas of this method are expanded continuously, and thus new applications are surfacing.

For example, the ability of the attached construct to bend precalibrated microneedles was used as a quantitative readout of the construct's response to known and unknown agonists for drug discovery [8], and the microneedles were employed as mechanical sensors in experiments with cardiomyocytes [79]. Moreover, the needles may be individually accessed and thus used either as electrical or biochemical sensors, or as drug delivery systems [80].

One other possible application of the Kenzan method deserves consideration: knowing the microneedles' position could help determine the initial distribution of the spheroids and of their cells (in the approximation of their semistochastic distribution), that is their "boundary conditions." In this way, a more precise examination of the basic biophysical mechanisms in arbitrarily shaped 3D constructs can be designed, which is difficult or even impossible to do otherwise.

Another extension of the Kenzan method could be to assist the biofabrication of tissues from multiple stem cells–derived (e.g., cardiac) organoids. These can be obtained in large quantities from induced pluripotent stem cells (iPSC), but unlike the adult cells–derived spheroids, their individual organization prevents their integration into a functional organ, usually remaining separated in the construct [81]. In this case, the Kenzan method may come to the rescue: these organoids could be selected in classes of equal sizes by sorting as larger particles [82], then assembled for long-term contact by the S-PIKE instrument in adjacent columns of variable diameters [83]. In addition, the molecular factors released by their needle piercing may actually trigger a "wound healing" reaction, further stimulating their fusion.

Last but not least, although the Kenzan method is by intent "scaffold-free," it could still be integrated with hydrogels in a "hybrid" bioprinting approach [3]. For example, alginate microbeads of the same size as the cell spheroids might be added to the constructs at the time of their assembling for local delivery of growth factors, as in [84].

6   Conclusions

The Kenzan method today stands out as a mature tissue engineering technology, capable to make large, complex, anatomically and physiologically meaningful tissue models. It fulfilled the dream of early tissue engineers to accomplish these goals "with cells only," although in a different way than other variants of bioprinting. As shown here, the success did not spring only from the ingenuity of this technology's inventors, but also from the mechanistic foundations of spheroids-based modular assembling, which can be explored in parallel both in experimental and in "virtual" modes.

Acknowledgments   The author is thankful to Drs. M. Swat and G. Oliveira and to P. Adhyapok for help with the computer model and to Dr. L. Moldovan for critical reading of the manuscript.

References

1. Moldovan NI, Hibino N, Nakayama K (2017) Principles of the Kenzan method for robotic cell spheroid-based three-dimensional bioprinting. Tissue Eng Part B Rev 23:237–244
2. Moldovan NI (2018) Progress in scaffold-free bioprinting for cardiovascular medicine. J Cell Mol Med 22:2964–2969
3. Moldovan NI, Moldovan L, Raghunath M (2019) Of balls, inks and cages: hybrid biofabrication of 3D tissue analogs. Int J Bioprint 5:167–175
4. Shimoto T, Nakayama K, Matsuda S, Iwamoto Y (2012) Building of HD MACs using cell processing robot for cartilage regeneration. J Rob Mechatronics 24:347–353
5. Itoh M, Nakayama K, Noguchi R, Kamohara K, Furukawa K, Uchihashi K, Toda S, Oyama J, Node K, Morita S (2015) Scaffold-free tubular tissues created by a bio-3D printer undergo remodeling and endothelialization when implanted in rat aortae. PLoS One 10:e0136681
6. Itoh M, Mukae Y, Kitsuka T, Arai K, Nakamura A, Uchihashi K, Toda S, Matsubayashi K, Oyama JI, Node K, Kami D, Gojo S, Morita S, Nishida T, Nakayama K, Kobayashi E (2019) Development of an immunodeficient pig model allowing long-term accommodation of artificial human vascular tubes. Nat Commun 10:2244
7. Ong CS, Fukunishi T, Zhang H, Huang CY, Nashed A, Blazeski A, DiSilvestre D, Vricella L, Conte J, Tung L, Tomaselli GF, Hibino N (2017) Biomaterial-free three-dimensional bioprinting of cardiac tissue using human induced pluripotent stem cell derived cardiomyocytes. Sci Rep 7:4566
8. Kitsuka T, Itoh M, Amamoto S, Arai KI, Oyama J, Node K, Toda S, Morita S, Nishida T, Nakayama K (2019) 2-Cl-C.OXT-A stimulates contraction through the suppression of phosphodiesterase activity in human induced pluripotent stem cell-derived cardiac organoids. PLoS One 14:e0213114
9. Arai K, Murata D, Verissimo AR, Mukae Y, Itoh M, Nakamura A, Morita S, Nakayama K (2018) Fabrication of scaffold-free tubular cardiac constructs using a bio-3D printer. PLoS One 13:e0209162
10. Machino R, Matsumoto K, Taura Y, Yamasaki N, Tagagi K, Tsuchiya T, Miyazaki T, Nakayama K, Nagayasu T (2015) Scaffold-free trachea tissue engineering using bioprinting. Am J Respir Crit Care Med 191:A5343
11. Yamamoto T, Funahashi Y, Mastukawa Y, Tsuji Y, Mizuno H, Nakayama K, Gotoh M (2015) Mp19-17 human urethra-engineered with human mesenchymal stem cell with maturation by

rearrangement of cells for self-organization – newly developed scaffold-free three-dimensional bio-printer. J Urol 193:e221–e222

12. Takeoka Y, Matsumoto K, Taniguchi D, Tsuchiya T, Machino R, Moriyama M, Oyama S, Tetsuo T, Taura Y, Takagi K, Yoshida T, Elgalad A, Matsuo N, Kunizaki M, Tobinaga S, Nonaka T, Hidaka S, Yamasaki N, Nakayama K, Nagayasu T (2019) Regeneration of esophagus using a scaffold-free biomimetic structure created with bio-three-dimensional printing. PLoS One 14:e0211339

13. Yanagi Y, Nakayama K, Taguchi T, Enosawa S, Tamura T, Yoshimaru K, Matsuura T, Hayashida M, Kohashi K, Oda Y, Yamaza T, Kobayashi E (2017) In vivo and ex vivo methods of growing a liver bud through tissue connection. Sci Rep 7:14085

14. Shafiee A, McCune M, Forgacs G, Kosztin I (2015) Post-deposition bioink self-assembly: a quantitative study. Biofabrication 7:045005

15. Groll J, Burdick JA, Cho DW, Derby B, Gelinsky M, Heilshorn SC, Jüngst T, Malda J, Mironov VA, Nakayama K, Ovsianikov A, Sun W, Takeuchi S, Yoo JJ, Woodfield TBF (2018) A definition of bioinks and their distinction from biomaterial inks. Biofabrication 11:013001

16. Sato S (2012) Ikebana: the art of arranging flowers. Tuttle Publishing, Tokyo

17. Kodama H (1981) A scheme for three-dimensional display by automatic fabrication of three-dimensional model. IEICE Trans Electron J64-C:237–241

18. Kodama H (1981) Automatic method for fabricating a three-dimensional plastic model with photo-hardening polymer. Rev Sci Instrum 52:1770–1773

19. Boland T, Mironov V, Gutowska A, Roth EA, Markwald RR (2003) Cell and organ printing 2: fusion of cell aggregates in three-dimensional gels. Anat Rec A Discov Mol Cell Evol Biol 272:497–502

20. Dutton G (2013) 3D printing may revolutionize drug R&D: organovo offers design options for modeling diverse tissue types. Genet Eng Biotechnol News 33:10, 12

21. Moldovan L, Babbey CM, Murphy MP, Moldovan NI (2017) Comparison of biomaterial-dependent and -independent bioprinting methods for cardiovascular medicine. Curr Opin Biomed Eng 2:124–131

22. LaBarge W, Morales A, Pretorius D, Kahn-Krell AM, Kannappan R, Zhang J (2019) Scaffold-free bioprinter utilizing layer-by-layer printing of cellular spheroids. Micromachines 10:570

23. Nguyen DG, Funk J, Robbins JB, Crogan-Grundy C, Presnell SC, Singer T, Roth AB (2016) Bioprinted 3D primary liver tissues allow assessment of organ-level response to clinical drug induced toxicity in vitro. PLoS One 11:e0158674

24. Langer EM, Allen-Petersen BL, King SM, Kendsersky ND, Turnidge MA, Kuziel GM, Riggers R, Samatham R, Amery TS, Jacques SL, Sheppard BC, Korkola JE, Muschler JL, Thibault G, Chang YH, Gray JW, Presnell SC, Nguyen DG, Sears RC (2019) Modeling tumor phenotypes in vitro with three-dimensional bioprinting. Cell Rep 26:608–623.e606

25. Moroni L, Boland T, Burdick JA, De MC, Derby B, Forgacs G, Groll J, Li Q, Malda J, Mironov VA, Mota C, Nakamura M, Shu W, Takeuchi S, TBF Woodfield TX, Yoo JJ, Vozzi G (2017) Biofabrication: a guide to technology and terminology. Trends Biotechnol 36(4):384–402

26. Groll J, Boland T, Blunk T, Burdick JA, Cho DW, Dalton PD, Derby B, Forgacs G, Li Q, Mironov VA, Moroni L, Nakamura M, Shu W, Takeuchi S, Vozzi G, Woodfield TB, Xu T, Yoo JJ, Malda J (2016) Biofabrication: reappraising the definition of an evolving field. Biofabrication 8:013001

27. Pourchet LJ, Thepot A, Albouy M, Courtial EJ, Boher A, Blum LJ, Marquette CA (2017) Human skin 3D bioprinting using scaffold-free approach. Adv Healthc Mater 6. https://doi.org/10.1002/adhm.201601101

28. Guo F, Mao Z, Chen Y, Xie Z, Lata JP, Li P, Ren L, Liu J, Yang J, Dao M, Suresh S, Huang TJ (2016) Three-dimensional manipulation of single cells using surface acoustic waves. Proc Natl Acad Sci U S A 113:1522–1527

29. Chen K, Wu M, Guo F, Li P, Chan CY, Mao Z, Li S, Ren L, Zhang R, Huang TJ (2016) Rapid formation of size-controllable multicellular spheroids via 3D acoustic tweezers. Lab Chip 16:2636–2643

30. Zhu Y, Serpooshan V, Wu S, Demirci U, Chen P, Guven S (2017) Tissue engineering of 3D organotypic microtissues by acoustic assembly. Methods Mol Biol 1576:301–312
31. Thomas J, Jones D, Moldovan L, Anghelina M, Gooch KJ, Moldovan NI (2018) Labeling of endothelial cells with magnetic microbeads by angiophagy. Biotechnol Lett 40(8):1189–1200
32. Mahajan KD, Nabar GM, Xue W, Anghelina M, Moldovan NI, Chalmers JJ, Winter JO (2017) Mechanotransduction effects on endothelial cell proliferation via CD31 and VEGFR2: implications for immunomagnetic separation. Biotechnol J 12. https://doi.org/10.1002/biot.201600750
33. Haisler WL, Timm DM, Gage JA, Tseng H, Killian TC, Souza GR (2013) Three-dimensional cell culturing by magnetic levitation. Nat Protoc 8:1940–1949
34. Tseng H, Gage JA, Haisler WL, Neeley SK, Shen T, Hebel C, Barthlow HG, Wagoner M, Souza GR (2016) A high-throughput in vitro ring assay for vasoactivity using magnetic 3D bioprinting. Sci Rep 6:30640
35. Parfenov VA, Koudan EV, Bulanova EA, PA Karalkin DASPF, Norkin NE, Knyazeva AD, Gryadunova AA, Petrov OF, Vasiliev MM, Myasnikov MI, Chernikov VP, Kasyanov VA, Marchenkov AY, Brakke K, Khesuani YD, Demirci U, Mironov VA (2018) Scaffold-free, label-free and nozzle-free biofabrication technology using magnetic levitational assembly. Biofabrication 10:034104
36. DB Solutions (2019) https://bioprinting.ru/en/press-center/publications/article-on-3dpmn-on-the-press-of-bone-on-board-the-iss/
37. Souza GR, Molina JR, Raphael RM, Ozawa MG, Stark DJ, Levin CS, Bronk LF, Ananta JS, Mandelin J, Georgescu MM, Bankson JA, Gelovani JG, Killian TC, Arap W, Pasqualini R (2010) Three-dimensional tissue culture based on magnetic cell levitation. Nat Nanotechnol 5:291–296
38. Olsen TR, Mattix B, Casco M, Herbst A, Williams C, Tarasidis A, Simionescu D, Visconti RP, Alexis F (2015) Manipulation of cellular spheroid composition and the effects on vascular tissue fusion. Acta Biomater 13:188 198
39. Tseng H, Gage JA, Shen T, Haisler WL, Neeley SK, Shiao S, Chen J, Desai PK, Liao A, Hebel C, Raphael RM, Becker JL, Souza GR (2015) A spheroid toxicity assay using magnetic 3D bioprinting and real-time mobile device-based imaging. Sci Rep 5:13987
40. Mironov V, Visconti RP, Kasyanov V, Forgacs G, Drake CJ, Markwald RR (2009) Organ printing: tissue spheroids as building blocks. Biomaterials 30:2164–2174
41. Yu Y, Moncal KK, Li J, Peng W, Rivero I, Martin JA, Ozbolat IT (2016) Three-dimensional bioprinting using self-assembling scalable scaffold-free "tissue strands" as a new bioink. Sci Rep 6:28714
42. Livoti CM, Morgan JR (2010) Self-assembly and tissue fusion of toroid-shaped minimal building units. Tissue Eng Part A 16:2051–2061
43. Blakely AM, Manning KL, Tripathi A, Morgan JR (2015) Bio-pick, place, and perfuse: a new instrument for three-dimensional tissue engineering. Tissue Eng Part C Methods 21:737–746
44. Gwyther TA, Hu JZ, Christakis AG, Skorinko JK, Shaw SM, Billiar KL, Rolle MW (2011) Engineered vascular tissue fabricated from aggregated smooth muscle cells. Cells Tissues Organs 194:13–24
45. van Pel DM, Harada K, Song D, Naus CC, Sin WC (2018) Modelling glioma invasion using 3D bioprinting and scaffold-free 3D culture. J Cell Commun Signal 12:723–730
46. Ovsianikov A, Khademhosseini A, Mironov V (2018) The synergy of scaffold-based and scaffold-free tissue engineering strategies. Trends Biotechnol 36:348–357
47. Mironov V (2005) The second international workshop on bioprinting, biopatterning and bioassembly. Expert Opin Biol Ther 5:1111–1115
48. Longo G, Montevil M (2012) The inert vs. the living state of matter: extended criticality, time geometry, anti-entropy – an overview. Front Physiol 3:39
49. Auffray C, Imbeaud S, Roux-Rouquié M, Hood L (2003) Self-organized living systems: conjunction of a stable organization with chaotic fluctuations in biological space-time. Philos Trans R Soc London, Ser A 361:1125–1139

50. Montevil M, Mossio M, Pocheville A, Longo G (2016) Theoretical principles for biology: variation. Prog Biophys Mol Biol 122:36–50
51. Vickerman MB, Keith PA, McKay TL, Gedeon DJ, Watanabe M, Montano M, Karunamuni G, Kaiser PK, Sears JE, Ebrahem Q, Ribita D, Hylton AG, Parsons-Wingerter P (2009) VESGEN 2D: automated, user-interactive software for quantification and mapping of angiogenic and lymphangiogenic trees and networks. Anat Rec (Hoboken) 292:320–332
52. Piedade AP (2019) 4D printing: the shape-morphing in additive manufacturing. J Funct Biomater 10:9
53. Ong CS, Nam L, Ong K, Krishnan A, Huang CY, Fukunishi T, Hibino N (2018) 3D and 4D bioprinting of the myocardium: current approaches, challenges, and future prospects. Biomed Res Int 2018:6497242
54. Pebworth M-P, Cismas SA, Asuri P (2014) A novel 2.5 D culture platform to investigate the role of stiffness gradients on adhesion-independent cell migration. PLoS One 9:e110453
55. Greiner AM, Richter B, Bastmeyer M (2012) Micro-engineered 3D scaffolds for cell culture studies. Macromol Biosci 12:1301–1314
56. H Consortium (2019) The human body at cellular resolution: the NIH Human Biomolecular Atlas Program. Nature 574:187
57. Weber GM, Ju Y, Börner K (2019) Considerations for using the vasculature as a coordinate system to map all the cells in the human body. arXiv preprint arXiv:1911.02995
58. Zadpoor AA, Malda J (2017) Additive manufacturing of biomaterials, tissues, and organs. Ann Biomed Eng 45:1–11
59. Choudhury D, Anand S, Naing MW (2018) The arrival of commercial bioprinters – towards 3D bioprinting revolution. Int J Bioprint 4:139
60. Vlahos AE, Cober N, Sefton MV (2017) Modular tissue engineering for the vascularization of subcutaneously transplanted pancreatic islets. Proc Natl Acad Sci 114:9337–9342
61. Rago AP, Chai PR, Morgan JR (2009) Encapsulated arrays of self-assembled microtissues: an alternative to spherical microcapsules. Tissue Eng Part A 15:387–395
62. Ip BC, Cui F, Tripathi A, Morgan JR (2016) The bio-gripper: a fluid-driven micro-manipulator of living tissue constructs for additive bio-manufacturing. Biofabrication 8:025015
63. Nycz CJ, Strobel HA, Suqui K, Grosha J, Fischer GS, Rolle MW (2019) A method for high-throughput robotic assembly of three-dimensional vascular tissue. Tissue Eng Part A 25:1251–1260
64. Neagu A (2017) Role of computer simulation to predict the outcome of 3D bioprinting. J 3D Print Med 1:103–121
65. Wang Y, Kim MH, Tabaei SR, Park JH, Na K, Chung S, Zhdanov VP, Cho N-J (2016) Spheroid formation of hepatocarcinoma cells in microwells: experiments and Monte Carlo simulations. PLoS One 11:e0161915
66. Robu A, Mironov V, Neagu A (2019) Using sacrificial cell spheroids for the bioprinting of perfusable 3D tissue and organ constructs: a computational study. Comput Math Methods Med 2019:7853586
67. Shafiee A, Ghadiri E, Williams D, Atala A (2019) Physics of cellular self-assembly – a microscopic model and mathematical framework for faster maturation of bioprinted tissues. Bioprinting 14:e00047
68. Swat MH, Thomas GL, Shirinifard A, Clendenon SG, Glazier JA (2015) Emergent stratification in solid tumors selects for reduced cohesion of tumor cells: a multi-cell, virtual-tissue model of tumor evolution using CompuCell3D. PLoS One 10:e0127972
69. Thomas GLM, Mironov V, Nagy-Mehez A, Mombach JCM (2014) Dynamics of cell aggregates fusion: experiments and simulations. Physica A 395:247–254
70. Graner F, Glazier JA (1992) Simulation of biological cell sorting using a two-dimensional extended Potts model. Phys Rev Lett 69:2013–2016
71. Swat MH, Thomas GL, Belmonte JM, Shirinifard A, Hmeljak D, Glazier JA (2012) Multi-scale modeling of tissues using CompuCell3D. Methods Cell Biol 110:325–366

72. Sego TJ, Kasacheuski U, Hauersperger D, Tovar A, Moldovan NI (2017) A heuristic computational model of basic cellular processes and oxygenation during spheroid-dependent biofabrication. Biofabrication 9:024104
73. Beysens DA, Forgacs G, Glazier JA (2000) Cell sorting is analogous to phase ordering in fluids. Proc Natl Acad Sci U S A 97:9467–9471
74. Cerchiari AE, Garbe JC, Jee NY, Todhunter ME, Broaders KE, Peehl DM, Desai TA, LaBarge MA, Thomson M, Gartner ZJ (2015) A strategy for tissue self-organization that is robust to cellular heterogeneity and plasticity. Proc Natl Acad Sci U S A 112:2287–2292
75. Buno KP, Chen X, Weibel JA, Thiede SN, Garimella SV, Yoder MC, Voytik-Harbin SL (2016) In vitro multitissue interface model supports rapid vasculogenesis and mechanistic study of vascularization across tissue compartments. ACS Appl Mater Interfaces 8:21848–21860
76. Moldovan L, Barnard A, Gil CH, Lin Y, Grant MB, Yoder MC, Prasain N, Moldovan NI (2017) iPSC-derived vascular cell spheroids as building blocks for scaffold-free biofabrication. Biotechnol J 12. https://doi.org/10.1002/biot.201700444
77. Tseng H, Gage JA, Raphael RM, Moore RH, Killian TC, Grande-Allen KJ, Souza GR (2013) Assembly of a three-dimensional multitype bronchiole coculture model using magnetic levitation. Tissue Eng Part C Methods 19:665–675
78. Robu A, Aldea R, Munteanu O, Neagu M, Stoicu-Tivadar L, Neagu A (2012) Computer simulations of in vitro morphogenesis. Biosystems 109:430–443
79. Mukae Y, Itoh M, Noguchi R, Furukawa K, Arai KI, Oyama JI, Toda S, Nakayama K, Node K, Morita S (2018) The addition of human iPS cell-derived neural progenitors changes the contraction of human iPS cell-derived cardiac spheroids. Tissue Cell 53:61–67
80. Faccini de Lima C, van der Elst LA, Koraganji VN, Zheng M, Gokce Kurtoglu M, Gumennik A (2019) Towards digital manufacturing of smart multimaterial fibers. Nanoscale Res Lett 14:209
81. Skylar-Scott MA, Uzel SGM, Nam LL, Ahrens JH, Truby RL, Damaraju S, Lewis JA (2019) Biomanufacturing of organ-specific tissues with high cellular density and embedded vascular channels. Sci Adv 5:eaaw2459
82. Gassmann K, Baumann J, Giersiefer S, Schuwald J, Schreiber T, Merk H, Fritsche E (2012) Automated neurosphere sorting and plating by the COPAS large particle sorter is a suitable method for high-throughput 3D in vitro applications. Toxicol In Vitro 26:993–1000
83. Xiang Y, Yoshiaki T, Patterson B, Cakir B, Kim KY, Cho YS, Park IH (2018) Generation and fusion of human cortical and medial ganglionic eminence brain organoids. Curr Protoc Stem Cell Biol 47:e61
84. Strobel HA, Dikina AD, Levi K, Solorio LD, Alsberg E, Rolle MW (2017) Cellular self-assembly with microsphere incorporation for growth factor delivery within engineered vascular tissue rings. Tissue Eng Part A 23:143–155

Development of the Bio 3D Printer "Regenova®," the Robotic System

Masahiro Sakamoto and Kenji Yoneda

Abstract A unique bio 3D printer has been created using a specific type of scaffolding that can be configured to produce unique tissue configurations. The work described led to the development of an automated bio 3D printer known by the commercial name Regenova, a comprehensively capable system which can be utilized in both research and commercial industrial tissue engineering applications. The chapter provides a step-by-step descriptions of (1) the research and developmental work done by Dr. Nakayama, (2) the functions of Regenova and the operation outlines, and (3) the system developed by Shibuya Corporation, which can consistently produce cell constructs from expansion cultures with an isolator system without any contamination risk.

Keywords Robotic system · Bio 3D printer · Spheroids · 3D cell tissue construct · *Kenzan*

1 Introduction

1.1 The First Encounter with Dr. Nakayama

On May 22, 2009, Kenji Yoneda of Shibuya Corporation (Shibuya) was introduced via Email to Dr. Koichi Nakayama by Hajime Ohgushi MD at Advanced Industrial Science and Technology in Kakogawa city, Hyogo.

Dr. Nakayama had already invented a method to create a centimetric cell construct for regenerative therapy by positioning and shaping submillimeter spheroids

M. Sakamoto · K. Yoneda (✉)
Shibuya Corporation, Kanazawa, Japan
e-mail: rm-info@shibuya.co.jp

© Springer Nature Switzerland AG 2021
K. Nakayama (ed.), *Kenzan Method for Scaffold-Free Biofabrication*,
https://doi.org/10.1007/978-3-030-58688-1_3

which contained thousands of cells without using biomaterials, later called the "*Kenzan* method."

When he got an idea of the *Kenzan* method in 2006, he simultaneously noticed the necessity and importance of robotic system to perform the *Kenzan* method because it required submillimeter fine maneuver, which is almost impossible by human hands.

Although he visited many Japanese hi-tech companies, mostly around his home town Fukuoka, asking support to develop the system, he could not find any company partner, mostly due to lack of realization possibility of the *Kenzan* method, inability to display future business opportunity, and lack of budgets for development.

Finally, he decided to learn computer programming to develop the prototype all by himself. At that time in 2009, he had already developed a prototype of the robotic system all by himself that could create the cell construct without contaminating biomaterials automatically by using a computer vision method. His prototype, named "NK-1," could detect the top of each needle of XY coordination, center coordination and diameter of the spheroids, capture the desired spheroids via PC screen, and then skewer the spheroid into the desired needle semiautomatically.

Since he is an orthopaedic surgeon, he never systematically learned computer science or hardware engineering; his prototype lacked accuracy and was far from future clinical application.

On June 5, 2006, Mr. Yoneda and his colleague visited Dr. Nakayama's lab at Fukuoka.

Mr. Yoneda and his colleague were surprised at what Dr. Nakayama had invented because Mr. Yoneda was taught that to make 3D cell tissue construct it is important to use biomaterials as scaffold.

Dr. Nakayama's method was quite unique in that it did not require mixing biomaterials as scaffold. They were also impressed with the prototype "NK-1," soon realizing that it required a lot of improvement.

Shibuya is the leading manufacturer of bottling systems in Japan and has a very high level of expertise in the aseptic manufacture of biological and drug products in compliance with all international regulations. These systems are based on highly advanced aseptic, robotic machine automation and vision technologies. These are world-class systems compliant with global Good Manufacturing Practice (GMP) requirements and validated to the highest international standards. In addition, Shibuya possesses a breadth of design, engineering, and manufacturing capabilities in the semiconductor, food/beverage, and medical device fields.

The skills Shibuya had could improve and sophisticate Dr. Nakayama's robotic system.

Fortuitously, Shibuya already had a team in place to evaluate whether the company should use the knowledge base and core skills they had developed in high technology manufacturing to enter the field of regenerative medicine. Therefore, Shibuya welcomed the technology collaboration with Dr. Nakayama.

1.2 Base Technology

Dr. Nakayama developed the prototype to evaluate the functionality of the automation system. This development project is described in detail in the section titled "History of the *Kenzan* method." The prototype successfully established the capability of the following basic technological developments.

1. The recognition of spheroids and the needle tip of the *Kenzan* using computer vision
2. Picking up the spheroids and placing onto the tip of the *Kenzan* needle
3. Proof of concept in the utilization of a Cartesian coordinate robot system

Dr. Nakayama constructed this prototype by acquiring and assembling the components himself. He also developed the required software, completed the conceptual as well as the finished functional design, and executed the performance testing, which verified system capability.

1.3 **SPHEROID BUILDER-1 (SB-1)**

On December 1, 2010, Dr. Nakayama visited Shibuya to see *SPHEROID BUILDER* Unit 1, developed by Shibuya. Initially, he thought that a vacuum system with a syringe pump would work for picking up spheroids; however, the device did not prove to work with sufficient reliability. Dr. Nakayama then requested the spheroid picking-up method be changed to puncture both sides of the spheroid with needles.

However, if this technique was applied, the processing capacity of the machine would be extremely low and would be unable to allow the production of a larger cell construct necessary for any clinical application. Additionally, the system would be difficult to automate to allow functionality at a commercial scale.

Therefore, we decided to develop a device that utilizes a picking system which can gently handle delicate cell constructs and move them precisely so that they can be centered and arrayed on extremely small diameter needles. To accomplish this the system had to accurately detect the top of each needle so that spheroids with cells could be arrayed on any needle configuration to an accuracy of a few micrometers.

Shibuya found the development of a reliable method for spheroid picking and placing challenging but finally was able to develop a negative pressure flow suction method using a vacuum pump and a buffer tank.

The development of a reliable pick and place system for spheroids on needles took a long time. As a result, the final design specifications for machines was created in July 2010, and the manufacturing of Unit 1 was completed about 6 months later (Fig. 1).

First printing by *SB-1*

- Cell type: rabbit adipose-derived stem cells (rADSCs)
- Cells per a Spheroid: 5×10^4
- Spheroid size: 450 to 500 μm in diameter
- Needle Diameter 170 μm
- Horizontal needle pitch: 400 μm
- Vertical needle pitch: 350 μm
- Spheroid configuration: $6 \times 6 \times 13$ spheroids

First cell construct

- Adjacent spheroids adhered by each other
- Approx.size: 2.5mm×2.5mm×4.5mm

Fig. 1 First cell construct created by *SB-1*

1.4 **Regenova**

On March 6, 2013, we gave the product name to *SB-1*, which was the developing name of the machine. The following were finalized by voting from developers, and *Regenova* was chosen after hearing of foreign people.

1. *Regenova*
2. *Cellon*
3. *Tsumiki*

2 **Overview of *Regenova* (Bio 3D Printer)**

Regenova can position spheroids onto the *Kenzan* and create cell constructs. Components and operations of it are described below.

Kenzan In this study, we created a special instrument to fix spheroids, named "*Kenzan*," which is derived from a tool used for Japanese flower arrangement which has a lot of needles aligned in one direction.

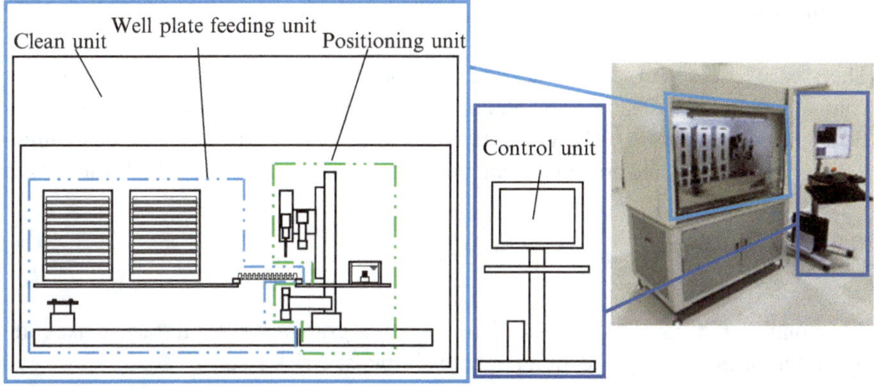

Fig. 2 Device configuration

2.1 Device Components

Regenova is mainly composed of four units: Fig. 2 shows the configuration of the device.

Positioning Unit

It is the most important unit for picking up spheroids and accurately positioning them onto the *Kenzan*. The unit consists of a pickup station that picks up spheroids, a station that organizes spheroids in preparation for moving them to the *Kenzan*, and a picking and positioning robot unit that handles spheroids and positions them onto the *Kenzan*. A camera unit is used by the picking and positioning robot to recognize the location of a spheroid and a suction nozzle so that the robot can pick up and move the spheroid. The camera unit can also guide the robot so that each spheroid can be precisely placed onto the *Kenzan*.

Well Plate Feeding Unit

Many spheroids are required to create a cell construct. Therefore, 96-well plates, on which the cells are grown on the spheroids, are fed from the stacking unit automatically. Once the spheroids are transferred to the *Kenzan*, the empty used well plates are transferred into the container unit where they are collected for a temporal storage. These transportation steps are conducted by the well plate transfer robot. This robotic unit can transport two types of spheroids, and the system can locate the appropriate well plate and transfer the specified spheroid.

Clean Unit

The area where spheroids are handled must be free of bacterial and foreign matter contamination. The clean unit is equipped with an air handler which supplies High Efficiency Particulate Air (HEPA)-filtered air in a vertical orientation. The unidirectional airflow is directed downward and bathes the work area in clean HEPA-filtered air.

Control Unit

The control unit consists of a PC. The installed graphic user interface makes the operation of the unit by a technician both simple and risk free.

2.2 Overview of Operations

In order to create a cell construct, a three-dimensional map is prepared to determine where spheroids should be placed on the *Kenzan*. The number of spheroids is calculated from this mapping data. Production personnel then prepare the necessary number of spheroids.

Figure 3 illustrates the workflow of the entire operation. (The inside of the frame indicates automatic positioning operations.) After the loading of 96-well plates and placement of the *Kenzan* within the *Regenova*, the operator activates the machine.

In the process operation workflow, first, the position of a spheroid in the well is confirmed by the camera, and next, each spheroid is picked up. During the pickup process, the camera ensures that the spheroid was properly removed from the well. Next, the positions of the needle tips of the *Kenzan* are recognized by the camera, and the spheroid is positioned on the *Kenzan* needle. When all spheroids on a well plate have been properly positioned, the now used well plate is removed and replaced with a new one, which allows the continuation of the spheroid placement process. By repeating this operation, the cell construct is created according to the *Kenzan* map program.

It takes approximately 10 s to pick up a single spheroid and to position it onto the *Kenzan*. Therefore, it takes about 16 min to position all of the spheroids on a single 96-well plate.

3 Required Function

There are a number of specialized systems which have been developed in order to accurately place spheroids on the *Kenzan*. The functional requirements of these systems are described below.

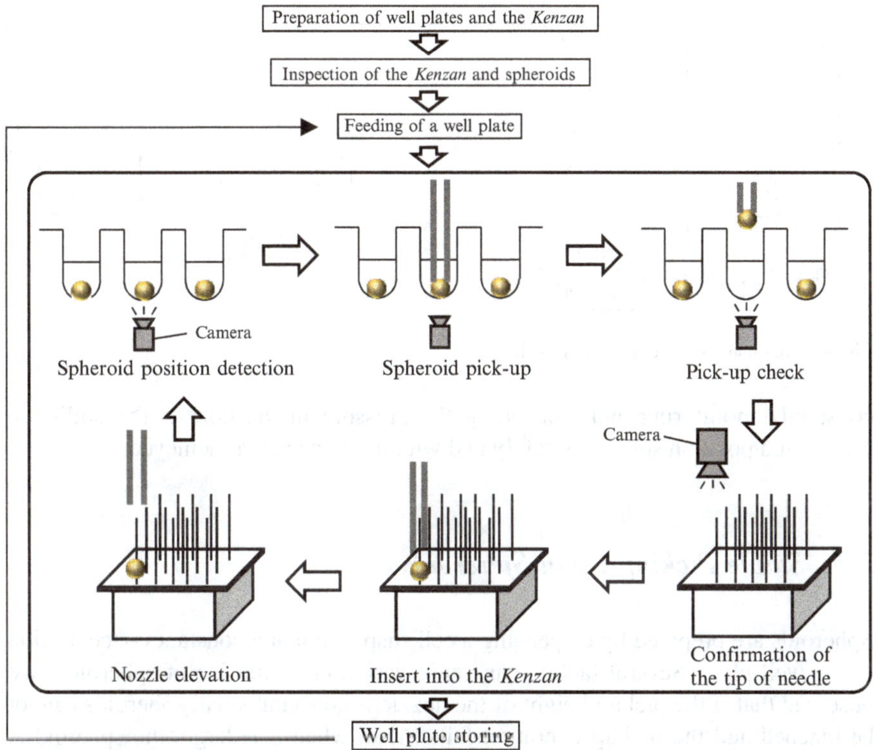

Fig. 3 Operation flow

3.1 Handling of Spheroids

In order to create a cell construct, it is necessary to be able to pick up a very small, fragile, and irregularly shaped spheroid and to move it to the *Kenzan*. We adopted a negative pressure flow suction method using a suction nozzle for handling spheroids. This handling step was developed for stable operation at high speed.

First, the specification of a suction nozzle is explained (Fig. 4). The inside of a suction needle must be hollow so that the needle of the *Kenzan* can be inserted into the cavity of the suction needle while spheroids are being picked up. Therefore, the inner diameter of the nozzle must be smaller than the diameter of spheroids but larger than the needle diameter of the *Kenzan*. Additionally, the outer diameter of the nozzle must be smaller than the diameter of spheroids because spheroids are positioned on the *Kenzan*.

The pressure maintained in the spheroid suction is also important, because if the suction force is too low, spheroid pickup is not possible. However, if the suction is strong, spheroids will be damaged or even ruined. Stable handling and prevention of damage to spheroids requires the precise maintenance of the proper pressure, but we found that pulsation occurs when a vacuum pump is used. In this equipment, a uniform suction without the pulsation was realized by providing a buffer section of the intermediate tank between the suction tube and the vacuum pump. In addition, by

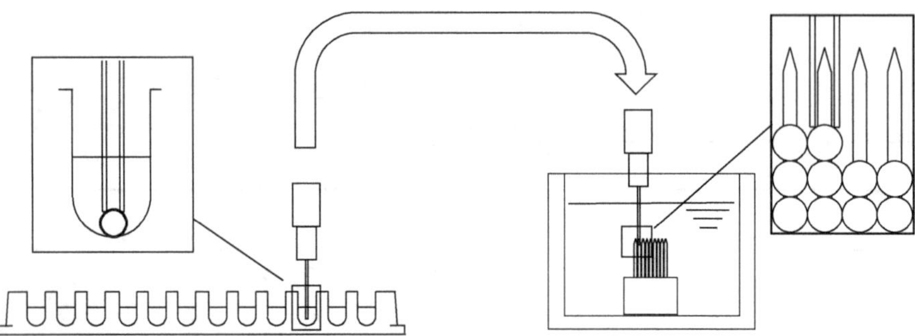

Fig. 4 Dimensions of the suction nozzle

constantly monitoring and controlling the pressure in the nozzle, the ability to pickup and position spheroids safely and without damage was achieved.

3.2 Stable Picking Up of Spheroids

Spheroids are prepared by dispensing a cell suspension at a constant concentration into a well plate. Several factors can cause variation in the size of spheroids. We observed that if the pickup height of the nozzle is constant, small spheroids cannot be reached and the pickup operation fails. Also, when it is large, the spheroid is squashed. To obviate these difficulties, a new algorithm using image processing was developed to enable fine adjustments in the pickup system to ensure operational efficiency and minimal damage to spheroids.

This algorithm measures the size of spheroids and calculates the pickup height of the nozzle when the position of the spheroid is recognized from the bottom of the well plate with a camera. As shown in Fig. 5A, the spheroid is almost spherical, which means the size of the bottom view is same as that of the side view, and the pickup height of the nozzle can be properly controlled by determining the size of the spheroid it is possible to adjust the nozzle position higher when the spheroid size is large and lower when it is small.

Another issue is that spheroids generally aggregate at the center of the wells, but some of them are occasionally located off-center. This necessitates a change in suction height depending on the location of the spheroid, even when it is the same in size. Thus, as shown in Fig. 5B, when the position of the spheroid is recognized by the camera, the proper height of the nozzle can be calculated from the shape (curve) of the well and inputted into the pick and position robot control system beforehand. Thus, the height of pickup can be controlled by making the nozzle lower when the spheroid is located in the center and making it higher according to the distance from the spheroid to the center. In addition, this algorithm is programmed so that when the size and shape of spheroids are examined and the spheroid does not meet the requirements for use, the spheroid is not picked and a spheroid meeting the required quality attributes is picked instead.

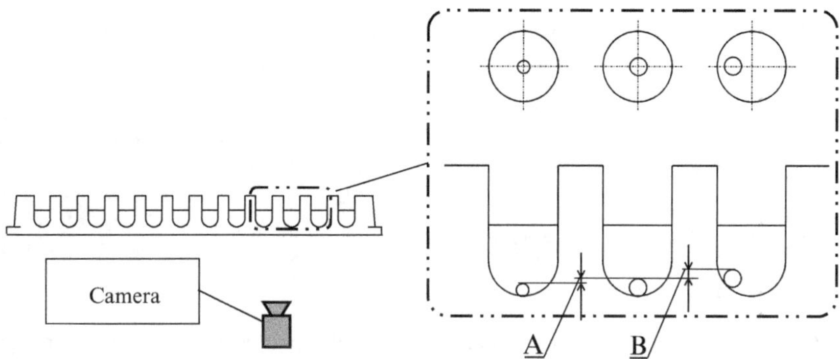

Fig. 5 Pickup height A schematic view of a system detecting the spheroid diameter and determining the pickup height. Spheroids are captured from the bottom of the well plate, and their diameters are calculated. (**A**) If a spheroid is small, the pickup height is controlled to lower. (**B**) If a spheroid is located off-center, it is controlled to higher.

However, even if picking is performed according to the characteristics of spheroids and well plates, it is not always possible to hold 100% of spheroids due to external factors such as clogging of the nozzle or the different characteristics of spheroids. Therefore, an inspection of each picked spheroid is performed as it is held immediately after picking. If a defect is detected, the picking operation is performed again. Additionally, if a picking failure occurs subsequently, for example if nozzle obstruction occurs, the nozzle is cleared with a puff of air and a discharge operation is performed to remove the obstruction.

These monitoring, control, and restoration functions allow a stable and high-speed picking operation to be realized.

3.3   *High-Precision Position Control by Image Processing*

In general, when high-precision position control is required, high-precision components are used; however, there is a limit to what can be achieved by hardware improvements alone. This is why the combined optimization of both hardware and software facilitated by image processing is very effective and essential to achieve the best results.

The suctioning nozzle used in *Regenova* should be a very fine, single-use product. Because single-use nozzles require frequent changes, it is very difficult account for minor fluctuations of the nozzle tips. Furthermore, a slight misalignment in the replacement of the nozzle is a fatal issue. To resolve these issues, a device was designed (see Fig. 6a) to detect the deflection width of the tip of the nozzle with a camera and to correct the position with software, thereby achieving high-precision repetitive position control.

In this correction control using the camera, the images taken greatly influence positional accuracy and the quality of the image is greatly influenced by the lighting which illuminates the photographic subject. Thus, in this equipment, suitable

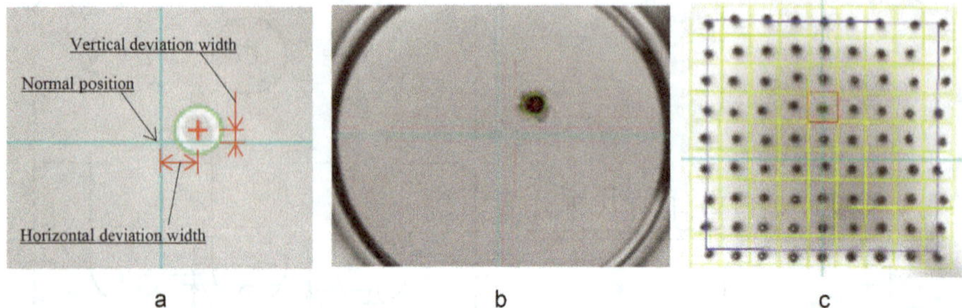

Fig. 6 Position recognition by image processing (**a**) nozzle tip recognition, (**b**) spheroid recognition, (**c**) the *Kenzan* needle recognition

lighting requirements for photographic examination subject were carefully determined and implemented. For example, a ring light was applied in the detection of the tip of the *Kenzan*, a backlight was applied in the detection of a spheroid, and a spotlight was applied in the detection of the tip of the nozzle. These selected lighting conditions yielded images of better quality, as shown in Fig. 6.

3.4 Positioning of Spheroids

The selection and placement of spheroids onto the *Kenzan* is described below. Spheroid characteristics vary depending on cell type and culture conditions. In order to realize a suitable quality construction, the following items were considered in developing the process and processing equipment:

Positioning Robot Operational Speed

The cycle time of the spheroid positioning robot governs the processing capacity of the equipment. In this step, the stability of the positioning operation is considered to be the most important factor and the robot's trajectory and speed are tuned to product optimal reliability and stability in operation. Tuning, including the characteristics of the cells and the suitability between cells and the *Kenzan*, is important as well as the speed and arrest time in spheroid picking and positioning. Also important is the stabilization time needed for image processing.

Pressure Control after Positioning

Another important matter during positioning is the internal pressure of the suction nozzle. When a spheroid is picked up, a negative pressure is generated in the inside of the nozzle and maintained even if the suction is stopped after positioning, with the result that the spheroid is picked up again from the *Kenzan*. We avoid this phenomena by introducing sterile compressed air into the suction line of the nozzle to break the negative pressure. The pressure control in the suction nozzle is also one of important factors for the pickup and positioning of spheroids.

Gapless Positioning

In order to produce a cell construct, the size of spheroids must be considered to ensure that spheroids adhere to each other and the correct position in pitch corresponding to their diameter is achieved. This equipment can change the pitch and the order of spheroid positioning to prevent any interference with the adjacent spheroids and to ensure that there are no gaps within the construction.

3.5 Modeling of Flexible Structures

The equipment must have the flexibility to position spheroids to enable cell constructs of various shapes to be produced. *Regenova* has a modeling function so that the positioning operation is conducted according to a preprogrammed design.

Many well-known 3D printers, 3D Computer-Aided Design (CAD), or CG software are typically used to create a construct model, but in this equipment, modeling is performed using a dedicated software which defines spheroid position along the needles of the *Kenzan*. For the first step of the modeling, the positioning pitch is determined and the mapping data to arrange spheroids on the *Kenzan* are created. An example of structural mapping data is shown in Fig. 7. The software can functionally rotate, zoom, and move the model and can highlight both the spheroids and the *Kenzan* needles in order to confirm the correctness of the map data. Also, spheroids can be displayed at any time during positioning to confirm progress. In addition, when two kinds of spheroids are used to build a cell construct, these

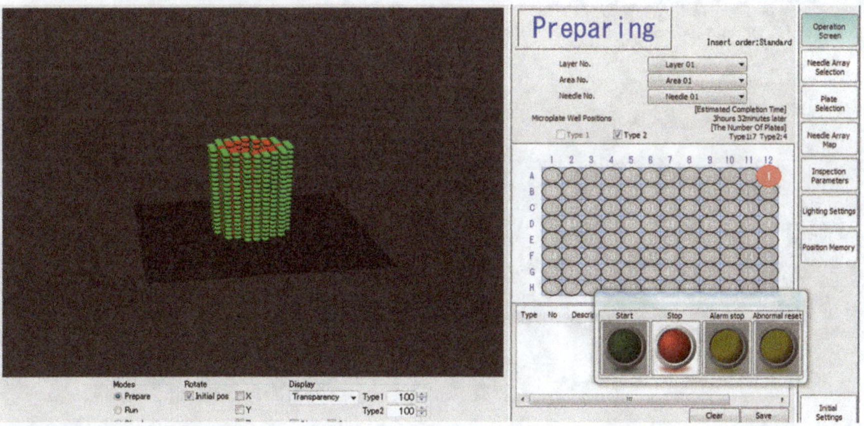

Fig. 7 Operation interface

spheroids can be color coded so they can be easily distinguished on the display. This feature makes it easy to understand the relative positioning of spheroids that contain different cells. These functions facilitate flexible modeling and allow the complete confirmation of a cell construct.

4 Cell Construct Manufacturing Systems for Clinical Applications

The maintenance of product sterility is a critical requirement that must be considered in the design of manufacturing equipment for clinical products.

In conventional cell processing facilities, it is common that operators wear an aseptic processing gown in the clean zone, often called the Cell Processing Center (CPC). These technicians conduct cell manipulation using aseptic techniques performed in a biological safety cabinet. However, there are a number of disadvantages of using a manned system, such as the one described above. First, the production area required is large and the running costs are correspondingly high. Second, the costs of the reusable or disposable aseptic gowns is high and they can adversely impact worker comfort. Instead of a manned CPC, isolator technology was applied for the production of a cell construct to provide the highest possible level of sterility assurance, reduce operating costs, and improve operator comfort.

There are several processes required to create a cell construct. If the functions of all processes are integrated into a single piece of equipment, the system must be fully occupied until each process step is completed. This requires a very high level of equipment flexibility and also results in deterioration of the operating rate of the equipment during the conduct of each process step.

Therefore, we developed a cell construct manufacturing system around a new concept, which is shown in Fig. 8, by creating a project for the "Development of

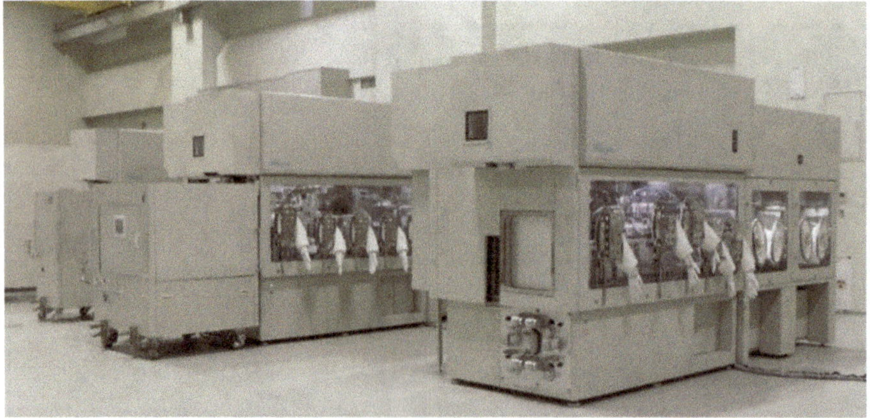

Fig. 8 Cell construct manufacturing system

Industrial Technologies for the Realization of Regenerative Medicine," which was commissioned by Okinawa Prefecture. In this development product, the manufacturing processes of the cell construct were divided into three steps:

1. Expansion culture of cells
2. Production of cell-containing spheroids
3. Positioning of the spheroids onto the *Kenzan*

According to the resulting conceptual design, the necessary equipment for each process was developed. The sterility assurance of the final cell construct product and efficient batch production were enabled by utilizing decontamination interface functions. Decontamination interfaces were used in the attachment of all necessary devices, such as an incubator, to the main cell processing isolator. Initially, the system was developed to be able to create only the smaller cell constructs, which is all that was required for proof-of-concept research studies. The larger cell constructs required for clinical applications could only be built after it was proven that the concept was feasible.

The three major equipment components and incubators that comprise the cell construct manufacturing system are described below:

4.1 Cell Culture Isolator

In order to grow the large population of cells needed to create a cell construct, cell culture operations are performed, including cell expansion and cell suspension preparation. The isolator technology system is utilized to ensure the maintenance of sterility throughout the process. The isolator work station is equipped with a centrifuge and a microscope, both of which are necessary for cell culture.

4.2 Spheroid Manufacturing Equipment

In this equipment, dispensing of cell suspensions to well plates, growth medium exchange, and a microscopic examination of spheroids are performed. Our development studies confirmed that this system results in the production of quality-ensured spheroids. These equipment operations use up to 384-well plates so that a sufficient number of spheroids can be produced for a large cell construct.

4.3 Bio 3D Printer

High-speed spheroid positioning is required for the production of large cell constructs. Therefore, we developed a system that could use up to four nozzle heads to pick up and position spheroids.

4.4 Incubator

The incubator can be connected to each of the three devices when necessary, and this can be accomplished while maintaining complete sterility assurance. In addition, two kinds of incubators, a small type for manual operation and a larger type with automatic functions for feeding and storing, can be used according to production requirements. This feature allows for the manufacturing of cell constructs from multiple samples.

The equipment developed through this process is now sufficient to manufacture clinical scale cell constructs with extremely high levels of patient safety owing to superb sterility assurance provided by isolator technology. In addition to safety, the equipment has the ability to operate efficiently and can create up to four cell constructs at a time.

5 Conclusion

As of July 2019, approximately 20 *Regenova* are in operation both in Japan and overseas around the world. Dr. Nakayama's research has also been dramatically improved by the automated workflow enabled by the process refinements which this project delivered. One project is currently in clinical evaluation, which we hope will lead to commercial approval.

Large numbers of cells are required to produce a clinical cell construct, which means that manual operations relying on conventional cell culture methods are not practical. The development of automated mass cell culture technology is an important advancement, but there are still a lot of challenges of improvements in efficiency so that many engineered tissue products can help patients in the near future.

We plan to further contribute to the development of regenerative medicine by utilizing the technologies developed so far, such as automation, sterilization, and validation, and by continuing to improve aseptic automated cell manufacturing systems. These systems with future developments to come will provide safe and affordable products to a number of patients.

In the future, transplantation of a construct made only of cells will be developed, which will hopefully lead to the transplantation of scaffold-free regenerative organ composed only of cells. Ongoing clinical research must be the first step toward the realization of these dreams. Wherever the future development of regenerative medicine takes us, we believe that the industrialization of these products is possible only through mechanization and automation using the latest technology. It is only through the application of new technologies that state-of-the-art regenerative medicines for all people can become a reality.

Development of a Compact Bio 3D Printer, "S-PIKE®"

Yasuto Kishii

Abstract Cell products consisting of human cells are being actively researched and developed in the areas of regenerative medicine and drug testing platforms. In regenerative medicine, some cell products are used as cell injections, while others are used as a graft with three-dimensional (3D) structures. Using the Kenzan method, it is possible to produce large 3D cell structures that are several centimeters in size. Cyfuse Biomedical K.K. (Cyfuse) is developing 3D cell products as well as equipment for Bio 3D printing. Cyfuse has developed a Bio 3D printer with the following features: (1) it is sufficiently compact to set up in a biosafety cabinet; (2) it allows the flexibility to fabricate several forms of multicellular constructs, in order to expand its potential contribution to research. This chapter introduces the concept of equipment for automatic manufacturing of cell products in regenerative medicine and the functions required for Bio 3D printing using the Kenzan method.

Keywords Bio 3D printing · Cell spheroids · Kenzan method ·
Contamination control

1 Introduction

In recent years, cell products have been developed for many industries including food, agriculture, energy, and medicine. Significantly, products consisting of human cells are being actively researched and developed for the field of regenerative medicine. In regenerative medicine, some of these products are used as cell injections, while others are used as grafts with three-dimensional (3D) structures. Research and development of 3D cell products manufactured using the Kenzan method are ongoing for

Y. Kishii (✉)
Cyfuse Biomedical K.K, University of Tokyo Entrepreneur Plaza, Tokyo, Japan
e-mail: yasuto.kishii@cyfusebm.com

© Springer Nature Switzerland AG 2021
K. Nakayama (ed.), *Kenzan Method for Scaffold-Free Biofabrication*,
https://doi.org/10.1007/978-3-030-58688-1_4

applications in regenerative medicine and drug testing. The Kenzan method involves producing 3D cell structures without scaffolds and molds; instead, exclusively the power of cell fusion is used. In particular, multicellular spheroids that are approximately 0.5 mm in size and composed of several tens of thousands of cells are aligned on the needle. The arrangement of spheroids is performed using dedicated software while they are in contact with each other. The spheroids fuse together during maturation and form the final 3D cell structure. Using the Kenzan method, it is possible to produce large 3D cell structures that are several centimeters in size; such structures are being researched and developed for some tissues and organs in regenerative medicine.

Cyfuse Biomedical K.K. (Cyfuse) is a biotechnology start-up in the field of regenerative medicine. Cyfuse is working in collaboration with multiple research institutions and universities on the practical applications of regenerative medicine for various tissues and organs, including the development of multicellular artificial blood vessels, nerve conduit tubes to promote regeneration of peripheral nerves, and liver-like 3D constructs for drug testing tools. While there has been progress in the development of 3D cell products for some functions, the Kenzan method is also being developed for applications in industrial fabrication, in order to facilitate the mass production of 3D cell products. Cyfuse is also developing industrial technology for the manufacturing of cell products in order to provide high-quality cell products to more people. The cell product manufacturing process is outlined in Fig. 1.

When scaling up the production process from the research and development stage to the manufacturing stage, consistent quality and enhanced productivity are required. Manufactured products should comply with good manufacturing practice

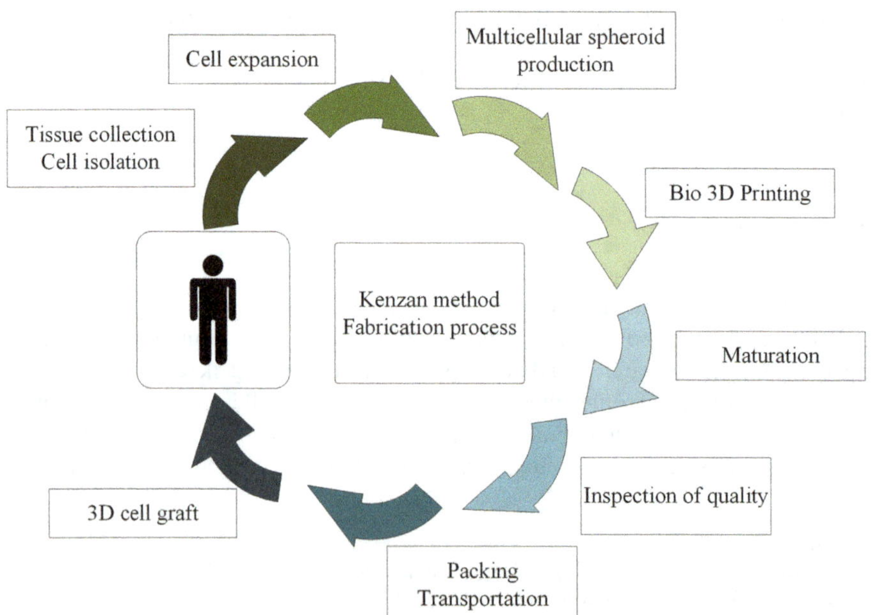

Fig. 1 Overview of cell product manufacturing process

(GMP) regulations of different countries to ensure quality, especially in manufacturing products for regenerative medicine. GMP regulations are rules concerning "manufacturing equipment," "manufacturing process management," and "quality control," to ensure that new products being manufactured have the same quality as established products. In particular, maintaining the quality of regenerative medicine products is essential to ensuring their safety and efficacy. Safety assurance means that products are manufactured in an environment where no extraneous materials are mixed because regenerative medicine products are administered to humans as therapy. This technique is called "contamination control." The maximum efficacy of established products can be achieved when medical professionals use the product properly under its intended use conditions. It is not possible to determine the equivalency of established products by quality standard alone because it is difficult to identify important quality characteristics that relate to the efficacy of regenerative medicine products used as therapeutics. The raw materials of regenerative medicine products are non-uniform and very sensitive to manufacturing processes. As such, maintaining consistent quality can be accomplished by managing raw materials and manufacturing processes. Most of the manufacturing processes for regenerative medicine include skillful manual operations. It is difficult to break expert operations down into basic operations, but automated and mechanized processes typically have higher repeatability, eliminate variation in quality due to differences in operator skills, and reduce the risk of foreign matter contamination. This chapter introduces approaches for contamination control in mechanization by discussing cell processing equipment for manufacturing using the Kenzan method. It also introduces some of the technological developments aimed at enhancing the productivity of the Bio 3D printing process.

2 Contamination Control Technology

Contamination control involves creating a sterile fabrication environment where no extraneous materials other than the essential raw materials are used; this is essential for safety assurance. Contamination control does not include impurities derived from raw materials. Manufacturing processes are managed according to the GMP regulations of each country. Buildings and facilities are classified according to aseptic processing requirements. A "critical area" refers to an area where aseptic processing operations are performed. "Direct support areas" refer to those areas meant for supporting aseptic processing operations. In particular, a critical area is where cells are handled directly in a biosafety cabinet or an isolator. Direct support areas are where production equipment is installed. The cleanliness of each area is monitored and managed according to two factors: "airborne particles" and "microorganisms in the environment." The control criteria for airborne particles are given in Table 1 based on local guideline references [1]. The clean area classifications provided by the International Organization for Standardization (ISO) are given in Table 2. Environmental microorganisms are monitored and managed according to two factors – namely, "airborne bacteria" and "falling bacteria." The acceptable standards for environmental microorganisms are also mentioned in the local guidelines and are presented in Table 3.

Table 1 Categories of clean areas, guidance on the manufacture of sterile pharmaceutical products by aseptic processing [1]

Area		Air cleanliness[a]	Maximum allowable number of airborne particles (particles/m^3)			
			Count under nonoperating conditions		Count under operating conditions	
			$\geq$ 0.5 µm	$\geq$ 5.0 µm	$\geq$ 0.5 µm	$\geq$ 5.0 µm
Aseptic processing area	Critical area	Grade A (ISO5)	3520	20	3520	20
	Direct support area	Grade B (ISO7)	3520	29	352,000	2900
Indirect support area		Grade C (ISO8)	352,000	2900	3520,000	29,000
		Grade D	3520,000	29,000	Dependent on process attributes[b]	

[a]The ISO class designation in parenthesis refers to the count during operation
[b]There are cases where maximum allowable number may not be specified

Table 2 ISO Classification of clean areas, ISO146441-1:2015

ISO classification number	Maximum concentration limit of particles (particles/m^3 of air)					
	Particle size					
	0.1 µm	0.2 µm	0.3 µm	0.5 µm	1 µm	5 µm
ISO class 1	10					
ISO class 2	100	24	10			
ISO class 3	1000	237	102	35		
ISO class 4	10,000	2370	1020	352	83	
ISO class 5	100,000	23,700	10,200	3520	832	
ISO class 6	1000,000	237,000	102,000	35,200	8320	293
ISO class 7				352,000	83,200	2930
ISO class 8				3520,000	832,000	29,300
ISO class 9				35,200,000	8320,000	293,000

Table 3 Acceptance criteria for environmental microorganism Count[a] (during operation), guidance on the manufacture of sterile pharmaceutical products by aseptic processing [1]

Cleanliness grade	Airborne microorganisms		Surface microorganisms	
	Air (CFU/ m^3)	Settle plate[b](CFU/ plate)	Contact plate (CFU/24– 30 cm^2)	Gloves (CFU/five fingers)
A	< 1	< 1	< 1	< 1
B	10	5	5	5
C	100	50	25	–
D	200	100	50	–

[a]Acceptance criteria are expressed as mean values
[b]Measurement time per plate is 4 hours at maximum, and the measurement is performed during processing operation

Cell processing equipment is typically set up in a direct support area, and cells are handled in the critical area inside the equipment. Therefore, during equipment design, it is necessary to consider contamination control for both areas. It is also important to account for the people involved in the entire life cycle of the equipment, from the setup and assembly to the operation and maintenance. There are three important principles concerning contamination control in the design of cell processing equipment:

1. Prevention: To prevent particle generation and growth of microorganisms inside equipment
2. Elimination: To remove particles and microorganisms from equipment
3. Protection: To block environmental contamination from entering the equipment

These principles are in detail in the sections that follow.

2.1   Prevention

Prevention is the first principle of contamination control and refers to preventing airborne particle generation and microorganism growth inside manufacturing equipment. During equipment operation, there are two main factors responsible for particle generation: sliding movement and thermal energy. Sliding movements cause grease dispersal and particle generation due to local pumping effects and the friction between cables and tubes. Mechanical movements inherently required involve sliding and bearing components. Methods for preventing particle generation due to sliding movements have been developed in the semiconductor industry and for microfabrication processes. Clean actuators are sold by many companies based on the ISO classification number. Particle generation can be further reduced by blocking dust dispersal with labyrinth structures so that the rotating and sliding parts of the processing equipment are not exposed and by cleaning the space with a vacuum pump. When designing sliding parts, the selection of driving parts is important, as is the layout of cables or tubes. Particles are generated too by thermal energy due to outgassing from electric wires or electronic devices, which is caused by temperature increases. Special cables have been developed in the semiconductor industry to suppress outgassing. In the case of heating equipment, appropriate materials must be selected. Plastic components that absorb considerable amounts of moisture generate significant outgassing with changes in temperature; therefore, plastic components should not be used for parts which are subjected to large temperature changes. It is possible to install the equipment (or a component of it) in a wind tunnel in order to measure particle generation during operation and evaluate equipment cleanliness. It is important to avoid selecting parts only for their extremely high cleanliness and to instead achieve the minimum necessary cleanliness standards for manufacturing by inspecting the cleanliness of each part, because excessive design and quality of components drive up cost of development and production for equipment.

It is also important to select appropriate materials for equipment parts and to consider the influence of airborne liquid droplets in order to suppress the growth of microorganisms. While metal and plastic materials generally do not pose problems, paper and wood should not be used in cell processing equipment. Care should be taken when handling liquids in cell processing equipment to maintain sterility. Airborne liquid droplets and dripping during liquid handling are major causes of cross-contamination. Thus, it is important to select appropriate pipette tips and pipetting methods during cell processing to avoid droplet production. Moreover, when trying to use up all the liquid in a bottle and the bottom of the container appears, it is essential to consider the potential generation of airborne droplets at the air–liquid interface.

2.2 Elimination

The second principle of contamination control is elimination, which refers to the removal of particles and microorganisms from and the prevention of their storage and accumulation on equipment. Even if dust generation and particle entry are suppressed, it is difficult to completely prevent contamination caused by regular operation and maintenance work. Elimination involves maintaining cleanliness by continuously and regularly removing particles and microorganisms in the critical area. There are three main factors in this method: (i) the ease of sanitizing and cleaning the equipment, (ii) the decontamination and sterilization process, and (iii) the design of the airflow through the equipment.

First, the sterile area should be a restricted space in the critical area to facilitate easier cleaning. If the cell processing equipment is installed in a biosafety cabinet, it is very difficult to clean the entirety of it. However, repeated equipment movements usually occur in this restricted space, so it is not necessary to clean the area to perform tasks other than cell processing. In particular, the difficult-to-reach areas are inside devices, including motors, cameras, and sensors, and the components which are upstream of the airflow in a biosafety cabinet. The areas of focus during cleaning should be determined based on reasonable decision making and risk assessments. Second, suitable mechanical structures with uniform surfaces and few openings are necessary to facilitate easier cleaning, which usually involves wiping the surface with ethanol or another disinfectant. When suitable structures are installed, the entire sterile area can be wiped easily with a reduced risk of missing the cleaning of some parts, as would be the case with openings or bumps in the surface. When using adhesive components such as seals and tapes, the parts or printed areas with these adhesive components need to be scratch-resistant. Peeling off seals and tapes should be avoided as waste accumulation and microorganism growth can occur on the areas underneath the adhesive. Finally, it is important to appropriately design the layout of cables, tubes, and pipes. If cables and tubes are in contact with the floor, dust can easily enter the clearance area between the cable and floor. Thus, to enable effective manual cleaning, it is necessary to lay cables and tubes at a sufficient distance from the cell processing surface.

There are two predominant methods for decontamination and sterilization: gas and ultraviolet (UV) light. The gas used for decontamination is generally peracetic acid or hydrogen peroxide. Selection of appropriate materials and sealing structures is important for effective gas decontamination. Since iron, copper, and brass are highly susceptible to corrosion by the gases used for decontamination, these metals should not be used in the cell processing equipment. Conversely, it is appropriate to use stainless steel, aluminum, and aluminum-containing alloys, as they are corrosion-resistant. However, copper and copper alloys cannot be eliminated from processing equipment entirely because they are used in many of the essential electrical components. One solution to this problem is masking to prevent the entry of contaminants into electrical components. Notably, since there are not many actuators, high-precision cameras, or sensors with the specifications necessary for gas decontamination on the market, it is difficult to select electrical components sterilizable by this method. The approach of masking each device enables a variety of options for the selection of electrical components; however, it is necessary to inspect the masking and perform risk assessment.

UV rays are classified into three types according to their wavelengths: UV-A (315 to 400 nm), UV-B (280 to 315 nm), and UV-C (200 to 280 nm). Among UV-C rays, those having a wavelength of approximately 250 to 265 nm have a strong sterilizing effect because they are absorbed by DNA. Mercury lamps are generally used as germicidal lamps in biosafety cabinets. However, these lamps are large, and their manufacture, import, export, use, and disposal involve stringent regulations according to the Minamata Convention on mercury signed by 92 countries. Therefore, light-emitting diodes (LEDs) emitting UV rays with a wavelength of 265 nm have been developed in recent years, and their market is gradually expanding. The UV LED developed for sterilization is very small and can be used for a variety of devices and equipment. With regard to the sterilization effect of a UV LED sterilization device, such as the one shown in Fig. 2, it has been shown that cells are killed by approximately 5 min of exposure to the UV radiation.

The design of air current systems usually involves the downflow method. Particles and microorganisms remain in the environment indefinitely when ventilation frequency is not defined. Cleanliness can be increased by having a higher ventilation frequency with clean air because particles and microorganisms are ejected or attenuated. However, this method is not necessarily required because the movement patterns of the equipment are repetitive. There are measures to cover local critical areas used for processing cells by using laminar flow to create an air barrier to minimize the impact of particles by generating airflow from the critical area to the devices which generate particles.

2.3 Protection

The third principle of contamination control is protection, which refers to blocking particles and microorganisms from entering cell processing areas and equipment. The simplest system is a closed isolator. It is possible to block the impact of the external environment by isolating critical areas used for processing cells. For equipment involving manual operation, it is necessary to install the equipment in a biosafety

Fig. 2 Small sterilization
device for UV irradiation

cabinet with an open system that meets the biosafety cabinet standards (NSF/ANSI 49-2007, EN12469, JIS K 3800: 2009, etc.) for sterility and airflow. However, manual operation is the most critical factor in the entry of particles and microorganisms into the cell processing environment. Manual operations include replacing disposable parts, supplying raw materials, regular maintenance activities, and repairing the equipment. Contamination can be minimized by reducing the number of manual operations, which can be achieved by improving usability of equipment. For example, if operators need to perform complicated procedures to exchange parts and supply raw materials, the contamination risk is increased. Improving usability of the cell processing equipment therefore enhances productivity and reduces the risk of contamination, which leads to overall improvements in product quality.

3   Productivity Enhancement in Bio 3D Printing

Cyfuse is developing 3D cell products as well as equipment for Bio 3D printing. There are two types of Bio 3D printing using the Kenzan method, which are "Regenova" and "S-PIKE." Regenova was jointly developed with Shibuya Corporation and can be used for both research and clinical purposes. Using Regenova, Cyfuse has been jointly developing cellular products for the regeneration of blood vessels and peripheral nerves with Saga University and Kyoto University, respectively. These projects will soon undergo clinical study. S-PIKE was originally developed by Cyfuse with the following features. First, it is sufficiently compact to set up in a biosafety cabinet. Second, it allows significant flexibility for the form of the multicellular construct to be fabricated, in order to expand the research potential of regenerative medicine cell products. Moreover, the basic steps of the S-PIKE sequence facilitate productivity enhancement for the future. To improve throughput, Bio 3D printing is currently under further development to increase the processing speed for each step and to ensure that steps progress in parallel as in a production line. The Bio 3D printing process can be summarized broadly in the following basic steps:

1. Detection and inspection of multicellular spheroids
2. Collecting spheroids
3. Arranging spheroids on needles
4. Arrangement of spheroids in 3D space

3.1 Detection and Inspection of Spheroids

Some of the methods for producing spheroids are the microplate method, the hanging drop method, and production using special bags. It is necessary for adjoining spheroids, which are set on needles, to be in contact each other when using the Kenzan method. Therefore, there are two important aspects regarding the shape of spheroids for manufacturing: minimal variation in size distribution and the roundness of the spheroids. Roundness is defined as the ratio of the inner circle and the circumscribed circle of the spheroid outline. A Bio 3D printer has to detect, inspect, and sort spheroids produced using the abovementioned methods. In the S-PIKE method, spheroids are first dispensed manually to the dedicated tray. Then the equipment can detect, inspect, and sort multiple spheroids simultaneously. This process is shown in Fig. 3. We are also studying methods for the homogeneous production of spheroids in bulk. If this method can be established, it will facilitate improved manual operations between the processes of producing spheroids and Bio 3D printing.

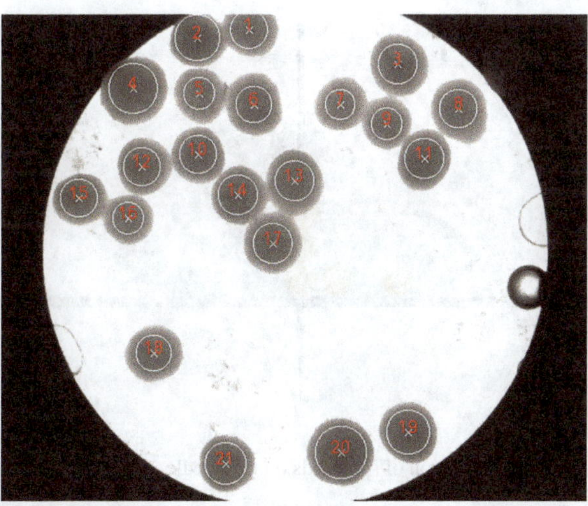

Fig. 3 Detection and inspection of spheroids on a dedicated tray

3.2   Collecting and Arranging Spheroids on Needles

After the detection and inspection of spheroids, the cell processing equipment must collect usable spheroids. There are many ways to accomplish this, but it is possible to maximize productivity by simultaneously collecting spheroids and placing them on needles. Specifically, the needles are operated by a robot, which picks up spheroids and places the selected spheroid directly on the needle. A depiction of this process is shown in Fig. 4. In arranging the spheroids, it is necessary to place them at appropriate positions on the needle. Therefore, the bottom face of the dedicated tray in which spheroids are dispensed is made of unwoven fabric, allowing the needle to pass through.

3.3   Arrangement of Spheroids in 3D Space

Spheroids are aligned on a needle at predetermined positions in 3D space. This is intended to position the needles in the desired orientation. Specifically, needles are aligned on a special silicone rubber platform atop a multilayered structure. Depending on the spheroid size, needles must be aligned so that adjoining spheroids are in contact. As the spheroids are produced using cells from sources that differ in race, age, and gender, it is very difficult to control spheroid size through process control. Therefore, a system that allows for adjustment of the arrangement of needles according to the size of the spheroids improves system usability and product quality. The arranged spheroids are matured on the needles and the 3D cell structure is subsequently produced. The alignment of spheroids to form a tube and an example of 3D cell structures are shown in Fig. 5.

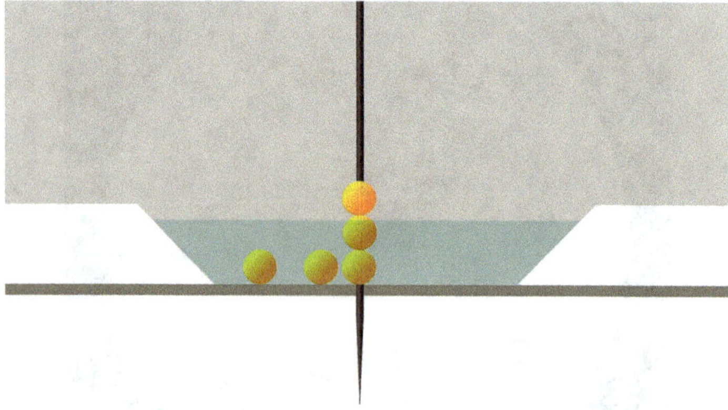

Fig. 4 A depiction of the collection of spheroids on the needle

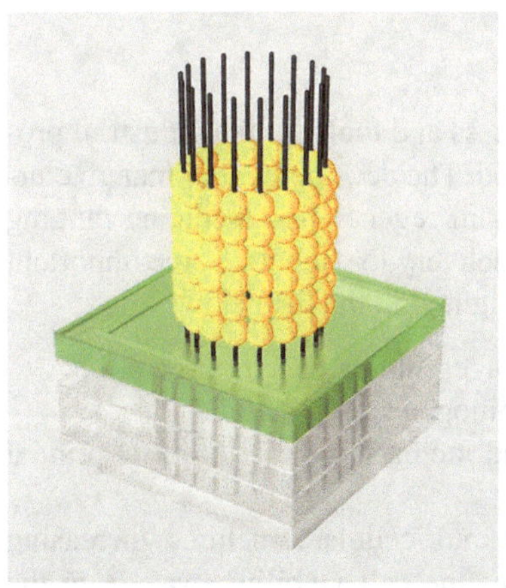  

Fig. 5 Arrangement and positioning of spheroids and a sample tubular 3D cell structure

Fig. 6 Bio 3D
printer: S-PIKE

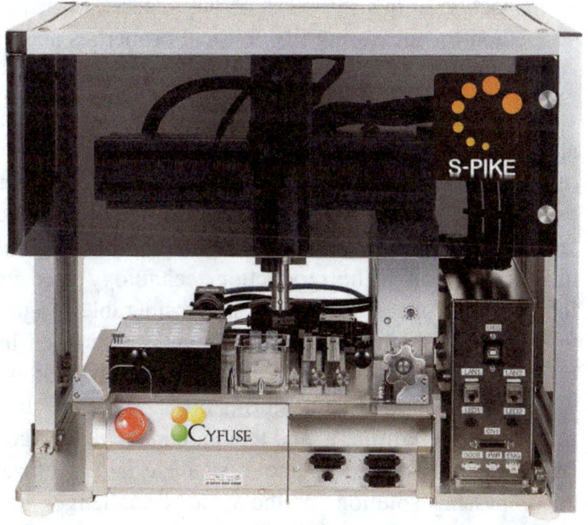

Steps 1 to 4 outlined in the introduction of Sect. 3 comprise the S-PIKE sequence. The S-PIKE of Bio 3D printer that is sold for research purposes is shown in Fig. 6. To enhance the bioprinter throughput, development is in progress to allow each step to be performed in parallel, as in a production line, and to construct a small-scale plant for production in a biosafety cabinet. This project is supported by the Tokyo Metropolitan Small and Medium Enterprise Support Center.

4   Future Prospects

This chapter described the Bio 3D printing process and tools for the industrial production of cell products using the Kenzan method. The development of manufacturing technology is ongoing for many processes in regenerative medicine ranging from tissue collection to product shipment and delivery to patients. Some important features of this manufacturing process are highlighted in the list below:

- Cell expansion – automation of medium changes and daily cell observation.
- Spheroid production – bulk production of homogeneous spheroids.
- Bio 3D printing – enhancing throughput and increasing the size of 3D-printed structures.
- Maturation – method for maturation of large, multicellular structures. Increasing the size of multicellular structures decreases the cell viability away from the surface of the structure. This is because the supply of oxygen and nutrients is insufficient past some distance from the surface, particularly in multicellular structures where capillary blood vessels are not developed.
- Inspection of quality – noncontact method for inspecting internal structures and method for measuring the physical properties and cell viability of multicellular structures.

It is required to ensure a consistent quality of manufacturing cell products at an appropriate cost, because regenerative medicine is leading edge but expensive. Most of the manufacturing processes for regenerative medicine are performed manually by skilled workers. However, automated and mechanized processes, in conjunction with skillful manual operations, are required throughout the regenerative medicine industry. Although the production technology used in the regenerative medicine industry is in its infancy, there is considerable progress in terms of international guidelines for materials and consumable parts used in product fabrication. If the fabrication process of cell products is standardized, it will lead to consistent product quality and reduced manufacturing costs. However, standardizing production of cell products is very difficult, and there are many challenges to be solved due to the different characteristics of the raw materials. Cyfuse aims to achieve an all-encompassing solution for the various challenges in cell product fabrication by working from both the biological perspective to develop the fabrication process and the engineering perspective to develop industrial technology. In the future, we believe that regenerative medicine will provide hope for people suffering from illness or injury.

References

1. Pharmaceuticals and Medical Devices Agency (2011) Guidance on the Manufacture of Sterile Pharmaceutical Products by Aseptic Processing. http://www.pmda.go.jp/english/review-services/gmp-qms-gctp/0001.html. Accessed 20 April 2011

Scaffold-Free Biofabrication for Articular Cartilage (and Subchondral Bone)

Daiki Murata

Abstract Osteoarthritis (OA) is a major joint disease that promotes locomotor deficiency with the associated disability and potentially decreases quality of life. In recent years, surgical methods to reconstruct both articular cartilage and subchondral bone for OA have been investigated to restore joint structure and function. In this chapter, several studies are introduced related to osteochondral regeneration using scaffold-free three-dimensional (3D) mesenchymal stem cell (MSC) constructs produced by using a "mold" and the "Kenzan" method. First, a study that demonstrated rabbit osteochondral regeneration following implantation of a "mold"-generated columnar construct consisting of autologous bone marrow-derived MSC spheroids is described. Second, examples are presented wherein the implantation using "mold"-made constructs consisted of autologous swine adipose tissue-derived (AT-) MSC spheroids deposited into osteochondral defects in minipigs. Next, a study concerning implantation of a "mold"-generated construct comprised of allogeneic AT-MSCs into a rabbit model is introduced. Furthermore, osteochondral regeneration following the autologous implantation of two swine AT-MSC tubular constructs fabricated by using the "Kenzan" method is considered, along with examples of histological examination of MSC constructs prepared by using either the "mold" or "Kenzan" methods. Finally, the future prospects of the "Kenzan" method for the further development of scaffold-free 3D cell constructions are discussed.

Keywords Articular cartilage · Mesenchymal stem cells · Osteoarthritis · Osteochondral regeneration · Scaffold-free

D. Murata (✉)
Center for Regenerative Medicine Research, Saga University, Saga, Japan

© Springer Nature Switzerland AG 2021
K. Nakayama (ed.), *Kenzan Method for Scaffold-Free Biofabrication*,
https://doi.org/10.1007/978-3-030-58688-1_5

1 Introduction

Articular cartilage constitutes hyaline cartilage, which generally contains no distribution of blood vessels or nerves and is composed of four layers including superficial, intermediate, radial, and calcified zones. The calcified zone provides ossification, with collagen fibers connecting the upper three layers to subchondral bone and articular cartilage being firmly attached to subchondral bone in epiphysis. Subchondral bone is comprised of cortical bone, which has abundant blood vessels and is engaged in supporting joint movement, protecting articular cartilage from bone axial load, and accelerating an articular cartilage metabolism. Although hyaline cartilage is generally encompassed by perichondrium consisting of collagen fibers and early mesenchymal progenitor cells that can differentiate into chondrocytes [1], articular cartilage is instead covered by horizontally arranged collagen-containing proteoglycans such as lubricin [2]. Therefore, the articular surface exhibits lubricity but does not possess self-healing ability owing to the absence of vascularity and perichondrium [3].

Osteochondral defects are frequently observed in young and active patients, with the repaired tissue being fibrous in nature because of the poor self-healing ability in articular cartilage as described above [4, 5]. Notably, this fibrous tissue does not have the functional properties of native hyaline cartilage. Therefore, the defects lead to osteochondral lesions, defined as cartilage degradation and subchondral bone sclerosis/deformity, and typically develop into osteoarthritis (OA) [6]. OA slowly progresses not only as a result of traumatic injuries to joint structures [7] but also through many exacerbating factors such as age, sex, body mass index, occupation, bone shape, and genetic factors regulating proteolytic enzymes [8–10]. In advanced OA, the progression of OA is also observed in response to cumulative damage to the bone, cartilage, and ligament, with cartilage degeneration and subchondral bone sclerosis potentially being worsened by the usual mechanical loads of daily activities [11]. Thus, OA constitutes a major joint disease that contributes to middle- and old-age locomotor deficiency [12–14], of which the associated disability can decrease quality of life. Therefore, surgical strategies to reconstruct both the bone and cartilage have been diligently investigated for restoring joint structure and function [12].

A particular focus in recent studies has been the complete regeneration of hyaline cartilage covering the subchondral bone. Recently, several treatments for damaged bone and cartilage including mosaicplasty [15], microfracture [16], and autologous chondrocyte implantation (ACI) [17] have been used in patients to relieve pain and improve joint function. Although a clinical study of mosaicplasty, which comprises the implantation of osteochondral autografts from non-load-bearing sites into the deteriorated (load-bearing) sites, showed favorable outcomes following surgery [13], it also demonstrated the inevitable loss of clinically sound cartilage [14], chondrocyte death associated with autologous osteochondral transfer [18], and limitations of available donor sites based on the required size and shape of the osteochondral autograft [14]. Moreover, the long-term results of mosaicplasty and microfracture are unsatisfactory because the replaced tissue does not constitute physiological hyaline cartilage but rather mostly fibrous cartilage, which lacks suitable mechanical properties [19]. Additionally, although multiple ACI technologies have already been commercialized and actually adapted to numerous patients [17], some previous studies on ACI have suggested the existence of associated problems such as isolation of only few chondrocytes from a small piece of normal cartilage [20], the small number of obtainable chondrocytes following in vitro culture [21], and dedifferentiation of chondrocytes during passaging in culture [22]. Therefore, with successive passaging over time, the potential exists for chondrocytes to convert to fibroblastic cells and cease providing hyaline cartilage-specific extracellular matrix (ECM) [23]. Finally, in histological evaluations of the repaired tissue, ACI exhibited no significant difference compared to microfracture in a randomized trial [24], with the repaired tissue in both cases comprised of predominantly fibrous cartilage.

To solve these problems, mesenchymal stem cells (MSCs) have recently received increasing attention as promising options for osteochondral regeneration [25, 26]. Stem cells are defined as immature cells that maintain the ability for self-renewal and the potential for multilineage differentiation into specific cells. In numerous previous studies, chondrogenic differentiation ability has been evaluated in vitro especially using MSCs derived from, e.g., bone marrow (BM), adipose tissue (AT), and synovial membrane (SM) [26–28]. BM-derived MSCs (BM-MSCs) have been mostly used to demonstrate differentiation into bone and cartilage in vitro [26]. Moreover, experimental implantation studies using BM-MSCs have been performed although the successful outcome of regeneration of hyaline cartilage and bone has been difficult to obtain in large animals [29–31]. In addition to AT-derived MSCs (AT-MSCs), ATs are recognized as an alternative source of BM-MSCs [27]. The advantages of AT-MSCs are that abundant cells can be isolated from AT and their cellular proliferation rate may be higher in mature animals [26, 32]. Although it has been reported that AT-MSCs rarely differentiate into chondrocytes [33, 34], AT-MSCs have been specifically shown to differentiate into cartilage in vitro [27]. Additionally, liposuction, which represents the surgical method used to aspirate AT, is commonly utilized in the cosmetic field and globally accepted for obtaining AT [35, 36]. In turn, SM-derived MSCs (SM-MSCs) have been reported to exhibit high chondrogenic potential in vitro [33, 34]; moreover, osteochondral regeneration using cell therapies has been evaluated in the knee of experimental animals using SM-MSC aggregates [37–39]. However, the conventional methods for transplanting cell suspensions have not been successful in reconstructing osteochondral defects in large animals [34]. Therefore, it is considered indispensable to develop methods for providing MSCs in a three-dimensional (3D) format.

In recent years, advances in tissue engineering of functional articular cartilage have been facilitated by several methods using synthetic or biological scaffolds to achieve sufficient thickness and mechanical function, along with supporting cell attachment, migration, proliferation, and differentiation [40, 41]. Accordingly, studies involving surgical procedures using a combination of artificial bone and autologous chondrocytes seeded into a collagen scaffold have also demonstrated favorable restoration of bone and cartilage compared to the outcomes of those using chondrocyte suspensions [20, 21]. Previous studies indicated that scaffolds composed of materials such as collagen and hyaluronic acid could be useful for promoting cell adhesion, proliferation, and chondrogenic differentiation [42, 43] and may also facilitate stem cell seeding for implantation into osteochondral defects [44, 45]. Bone regeneration using AT-MSCs seeded into hydroxyapatite has also been investigated [46]. However, artificial materials may induce xenobiotic reactions through immune response in the tissue [47] and may not support the completion of osteochondral regeneration [48, 49]. Notably, bone itself is self-restorative because its vascularity and cellularity allow ready degradation of the artificial bone (osteoclastic system) and subsequent reformation of new bone matrix (osteoblastic system) [3, 50]. In contrast, the articular cartilage has been suggested to be less restorative because it is formed only by chondrocytes, which both degrade and regenerate the special matrix [3, 50]. Furthermore, the collagen scaffold remaining in the implanted sites for long periods can prevent the regeneration of hyaline cartilage in addition to promoting its replacement to fibrous cartilage [49]. Accordingly, although numerous studies have reported the successful use of various biomaterials for scaffold construction, no ideal biomaterial has yet been identified. Considering the potential influence of scaffolds on the surrounding microenvironment [37], various concerns remain that need to be solved including immunogenicity [51, 52], long-term safety of scaffold degradation products [53], and risk of infection or transmission of disease [54]. Based on these concerns, it is possible that artificial scaffolds may in principle be unfavorable for regenerating articular cartilage; therefore, further investigation into the use of scaffold-free cell constructs to promote cartilage regeneration is necessary.

Toward this end, several scaffold-free systems have recently been investigated [55–57]; nevertheless, it has been difficult to create sufficient thickness to fill in the osteochondral defects without using a scaffold. For example, a thickness of 5 mm at least is necessary to fill a full-thickness articular cartilage defect in the knee. Furthermore, with or without a scaffold, it is even more challenging for the current methods to regenerate both articular cartilage and subchondral bone in the context of an osteochondral defect. Alternatively, it has been demonstrated that dissociated adherent cells grown in suspension have the capacity to form aggregates (spheroids) in nature to avoid cell death through cell-to-cell attachment [49]. The majority of the approaches using these spheroids incorporate a specific cylindrical mold to produce constructs of desired shapes [58, 59]. Moreover, certain cell types, such as fibroblasts and chondrocytes, possess the capacity to synthesize and release components of the ECM, such as collagen and proteoglycans, in vitro. Notably, this capacity is accelerated under confluence or 3D culture conditions because the cell cycle stops and the cells start to produce ECM components. As this phenomenon has already been confirmed in studies using MSCs, a novel method for fusing the spheroids by using a specific cylindrical "mold" was developed to fabricate scaffold-free 3D constructs consisting of MSCs in vitro [60–63]. At the bottom of the "mold," a base disc is placed into the chamber to facilitate handling of the construct (Fig. 1a). Therefore, loading the spheroids into the "mold" chamber can produce a columnar construct consisting of fused spheroids (Fig. 1b–e).

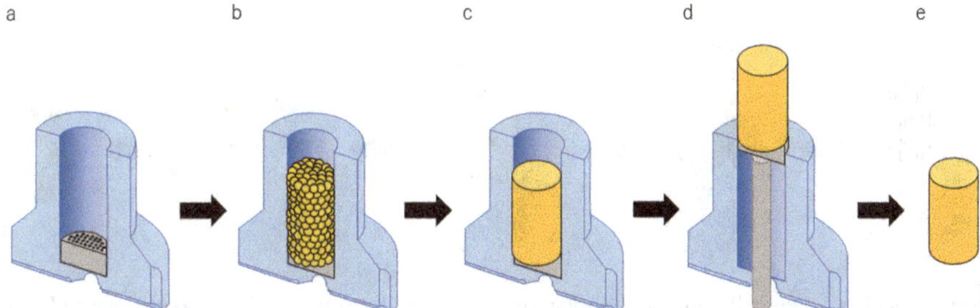

Fig. 1 Scaffold-free cell construct fabrication using the "mold" method. In this figure, half of the mold is illustrated for convenience of explanation. (**a–e**) Spheroids are employed to fabricate scaffold-free columnar cell constructs. (**a**) A cylindrical "mold" is prepared for fabricating a columnar construct. (**b**) Spheroids are piled up into a "mold." (**c**) Spheroids are cultured in the "mold" to fuse with each other. (**d**) The scaffold-free cell construct is retrieved from the "mold." (**e**) Cell constructs are used for implantation and analysis

Initially, the regeneration of bone and cartilage was demonstrated following implantation of scaffold-free 3D constructs consisting of autologous BM-MSC spheroids into osteochondral defects in the knee joints of rabbits [60]. However, pigs are expected to serve as more suitable animal models than rabbits for obtaining results that are more appropriate for extrapolating bone and cartilage regeneration to human OA. In particular, minipigs exhibit similar behavior patterns to human daily life, as they spend time standing and walking in the daytime and sleeping at night [64, 65]. Therefore, two studies have already been reported that were designed to evaluate the regeneration of articular cartilage and subchondral bone using scaffold-free columnar constructs consisting of autologous AT-MSC spheroids in minipigs [61, 62]. Nevertheless, these studies only evaluated the outcome of using autologous AT-MSCs for osteochondral regeneration [66, 67]. In contrast, the use of allogeneic AT-MSCs has not yet been reported, although previous studies showed the safety of the immune reaction and oncogenesis risk when using allogeneic MSCs and these cells offer some advantages including avoiding donor site morbidity and reducing overall cost [68, 69]. Consequently, data related to whether osteochondral defects could be healed histologically by implanting 3D cell constructs made of allogeneic AT-MSCs engrafted into osteochondral defects are only available in rabbit models.

However, "molded" constructs are often fragile and can be difficult to handle during surgical implantation, with the possibility of postsurgical defluxion. It has been suggested that construct fragility may be due to a shortage of ECM within the construct or to reduced perfusion of culture medium among the spheroids piled up in the cylindrical "mold," which would result in an uneven supply of oxygen and nutrients, in turn affecting ECM production and coalescence of the spheroids. Accordingly, to create more sophisticated constructs, the use of automatic spheroid deposition using bio-3D printing with a specific needle array (Fig. 2a), termed the "Kenzan" method, is recommended [70]. Bio-3D printing constitutes a technology to create tissues and organs with an internal structure by 3D-stacking spheroids at

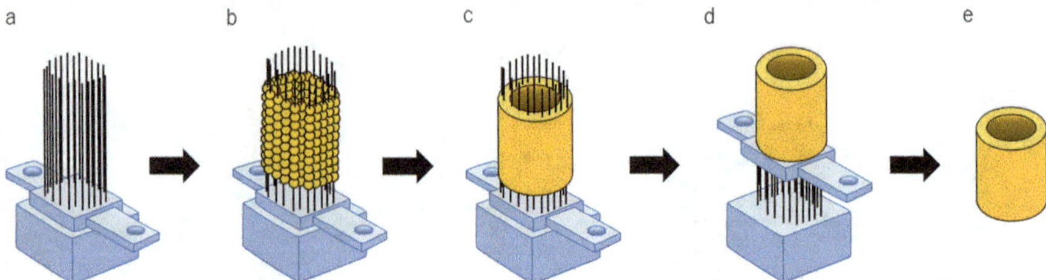

Fig. 2 Scaffold-free cell construct fabrication using the "Kenzan" method. (**a–e**) Spheroids are employed to fabricate scaffold-free tubular cell constructs. (**a**) "Kenzan" grids are prepared for bio-3D printing. (**b**) Spheroids are skewered onto the "Kenzan" automatically using a bio-3D printer. (**c**) Spheroids are cultured on the microneedles of the "Kenzan" to fuse with each other. (**d**) The scaffold-free cell construct is retrieved from the "Kenzan." (**e**) Cell constructs are cultured on tubular support for further maturation

appropriate positions of the array according to the original design. The "Kenzan," which plays the role of a temporary artificial scaffold to fix the position of the spheroids placed by bio-3D printing, is removed prior to the implantation after the spheroids fuse with each other using the increased ECM (Fig. 2b–e). With this technique, spheroids are uniformly dispensed in contact with one another at a regular distance and spacing, providing a more favorable condition for culture medium perfusion between the spheroids in vitro, thereby maintaining ECM production and cell viability. The bio-3D printing system can precisely dispense spheroids according to the size and shape of the osteochondral defect of interest. Therefore, it can be considered that maintaining ECM production and cell viability would likely contribute to the increased strength and elasticity of the construct and improve the tissue regeneration by also preventing postsurgical defluxion of the construct. Toward this end, a study was performed to evaluate the histology of AT-MSC constructs prepared using bio-3D printing using the "Kenzan" method and to analyze osteochondral regeneration following the autologous implantation of a dual construct composed of two AT-MSC constructs of different sizes into the site of osteochondral defect regeneration [71].

In this chapter, a study that demonstrated osteochondral regeneration following implantation of constructs consisting of autologous rabbit BM-MSC spheroids generated by using the mold is first introduced [60]. Second, examples are also mentioned that describe osteochondral regeneration through the implantation of constructs consisting of autologous swine AT-MSCs fabricated using the "mold" into osteochondral defects in minipigs [61, 62]. Subsequently, a study regarding a construct comprised of allogeneic AT-MSCs engrafted into osteochondral defects in rabbit models is presented [63], in addition to examples of histological examination for MSC constructs prepared by using the "mold" method [60, 63]. Finally, a representative study is indicated that evaluated the histology of AT-MSC constructs prepared using bio-3D printing using the Kenzan method and analyzed osteochondral regeneration following the autologous implantation of two swine AT-MSC constructs [71].

2 Osteochondral Regeneration Using Scaffold-Free 3D Constructs Produced by the "Mold" Method

2.1 Rabbit Osteochondral Regeneration with Autologous BM-MSCs

In this section, a study is introduced that demonstrates osteochondral regeneration following implantation of constructs consisting of autologous rabbit BM-MSC spheroids generated using the "mold" method [60]. Skeletally mature female Japanese white rabbits were used, from which bone marrow was aspirated from the iliac crest and the buffy coat of the bone marrow liquid was suctioned and expanded onto a culture dish in culture medium consisting of Dulbecco's modified Eagle's medium (DMEM) and 10% fetal bovine serum (FBS) (Table 1). BM-MSCs were isolated and expanded until they reached the necessary numbers. The collected BM-MSCs were divided into low-attachment 96 well plates at 4×10^4 cells/well (Table 1). After incubation for 2 days, the cells spontaneously formed spheroids (Table 1). Then, 600–700 spheroids were loaded into a specific cylindrical "mold" with a diameter of 4.6 mm (Table 1; Fig. 1a, b). The mold was maintained under air-liquid interface conditions and incubated for 1 week in the culture medium. These loaded spheroids fused with each other, resulting in the formation of a columnar construct that conformed to the inner shape of the "mold" (Fig. 1c–e). Following general anesthesia, a cylindrical defect was then created in the articular cartilage of the patella groove of the recipient rabbits (Fig. 3a). The size of the defect was set to 4.8 mm in diameter and 4–5 mm in depth. The depth was determined according to the height of the cell construct, although all reached the middle layer of the subchondral bone. The construct was gently implanted into the defect, and the top of the construct was flattened using the surgeon's finger (Fig. 3b). The osteochondral defect in the contralateral knee was left empty as a control. For obtaining the results of this study, individual rabbits were euthanized at 3, 6, 12, 24, and 52 weeks to confirm the osteochondral regeneration process by microcomputed tomography (micro-CT) analysis, in addition to macroscopic and microscopic (histological) evaluation.

Micro-CT scanning suggested that bone formations were present at the peripheral areas and developed toward the central area at 6 weeks following implantation. Macroscopic analysis revealed that the surface of the defect was covered with cartilaginous white tissue. Histological examination suggested that cartilaginous differentiation occurred at peripheral areas of the implanted constructs and developed toward the central area. After 12 weeks of implantation, endochondral bone formation progressed and almost the whole area corresponding to the subchondral bone was replaced by bone structures (Fig. 3c). In the joint surface area of the implanted tissue, hyaline cartilage was regenerated with a constant thickness (Fig. 3c). In contrast, in a control rabbit without any implantation, the defect was filled with fibrous tissue, and the cartilage formation and the subchondral bone formation were incomplete at the central area even at 12 weeks after implantation. Furthermore, micro-CT

Table 1 Fabrication conditions for the bio-3D constructs for in vivo differentiation

Animal [reference]	Cell source	Cell number of a spheroid/ spheroid size	Mold or Kenzan shape	Construct shape/ size	Medium	Culture period for spheroid / construct
Rabbit [60]	Autologous BM-MSCs	4.0×10^4 cells	Cylindrical mold (D; 4.6 mm)	Columnar	DMEM + 10% FBS	2 days
		N.D.		D; 4.6 mm H; 4-5 mm		1 week
MMPig [61]	Autologous AT-MSCs	5.0×10^4 cells	Cylindrical mold (D; 4.0 mm)	Columnar	DMEM + 10% FBS	2 days
		Approximately 700 μm		D; 4.0 mm H; 6.0 mm		1 week
Minipig [62]	Autologous AT-MSCs	5.0×10^4 cells	Cylindrical mold (D; 5.0 mm)	Columnar	DMEM + 10% FBS	2 days
		Approximately 500 μm		D; 5.0 mm H; 5.0 mm		7 weeks
Rabbit [63]	Allogenic AT-MSCs	5.0×10^4 cells	Cylindrical mold (D; 4.6 mm)	Columnar	DMEM +10% FBS	2 days
		Approximately 700 μm		D; 4.6 mm H; 3.0 mm		Several days
Minipig [71]	Autologous AT-MSCs	1.0×10^4 cells	Circular Kenzan (13 × 13)	Tubular	Combined medium[a]	1 day
				T; 1.5 mm iD; 2.0 mm H; 4.0 mm		
		Approximately 550 μm	Circular Kenzan (9 × 9)	Tubular		1 week
				T; 1.5 mm iD; 0.5 mm H; 4.0 mm		

[a]Combined culture medium; xeno-free MSC culture medium and serum-free MSC culture medium with a ratio of 1:1

MMPig microminipig, *BM-MSCs* bone marrow-derived mesenchymal stem cells, *AT-MSCs* adipose tissue-derived mesenchymal stem cells, *D* diameter, *H* height, *T* thickness, *iD* inner diameter, *DMEM* Dulbecco's modified Eagle's medium, *FBS* fetal bovine serum, *N.D.* no data

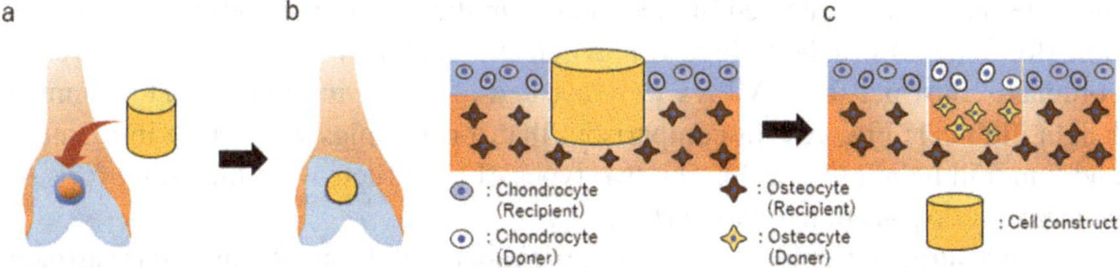

Fig. 3 Scaffold-free cell constructs produced by using the "mold" method for osteochondral regeneration in vivo. (**a**) Scaffold-free cell construct was prepared for the implantation. (**b**) Scaffold-free cell construct was implanted into an osteochondral defect. (**c**) Mesenchymal stem cells in the construct differentiated to chondrocytes and osteocytes, followed by regeneration of articular cartilage in the surface layer and formation of subchondral bone in the deep layer at the implanted site

scans revealed complete regeneration of the subchondral bone. Histological examination revealed that hyaline cartilage was maintained at a consistent thickness identical to that of the surrounding original cartilage at both 24 and 52 weeks in the treated defects (Fig. 3c). Tidemark, which separates hyaline cartilage from the calcified cartilage layer, was observed at these two time points, suggesting the integrity of the regenerated articular structures. In contrast, the defect was mainly filled with fibrous tissue or the formation of subchondral bone without tidemark in a control animal.

2.2 Swine Osteochondral Regeneration with Autologous AT-MSCs

In this section, examples that describe osteochondral regeneration following implantation of constructs consisting of autologous swine AT-MSC spheroids fabricated by using the "mold" are discussed [61, 62]. Skeletally mature microminipigs (MMPigs) and minipigs were used for these studies (Table 1). For obtaining cervical subcutaneous ATs of MMPigs, liposuction was performed (Table 1). Alternatively, liposuction was conducted for harvesting gluteal subcutaneous ATs of minipigs (Table 1). These AT samples were digested with proteolytic enzyme, and the collected cell suspensions were filtered and centrifuged. The pellets were resuspended and plated onto culture dishes or flasks in culture medium consisting of DMEM and 10% FBS (Table 1). Following incubation for 7 days, the cells adhering to the bottom of the dishes were harvested as swine AT-MSCs and cultured until construct creation. At least 4×10^7 AT-MSCs were used to produce each autologous construct for both MMPigs and minipigs. The cells were inoculated into eight low-attachment 96 well plates with 5×10^4 cells/well (Table 1). Following incubation for 48 h, the cells formed spheroids with a diameter of approximately 700 μm in the bottom of the wells (Table 1). About 760 spheroids were placed into a cylindrical "mold" (4 or 5 mm in diameter for MMPigs or minipigs, respectively) (Table 1; Fig. 1a, b). Then, the spheroids were incubated in the "molds" in the culture medium until implantation (for 7 days) (Table 1; Fig. 1c). When the "mold" was carefully removed, a columnar construct for MMPigs was revealed of 4 mm in diameter and 6 mm in height (Table 1; Fig. 1d). In comparison, that for minipigs was 5 mm in diameter and 5 mm in height (Table 1). The two types of constructs were then used for subsequent autologous implantation (Fig. 1e).

Using a surgical trephine with an outer diameter of 4 mm, the articular cartilage and subchondral bone were drilled to a depth of 6 mm at the center of the groove in MMPigs. After removing a column of cartilage and bone, a cylindrical osteochondral defect was created in each groove. The columnar construct was autografted into the osteochondral defect in the right hind limb (Fig. 3a, b), whereas no graft was implanted into the defect in the left limb (control defects). Alternatively, the surgery was performed under general anesthesia to create a cylindrical osteochondral defect (5 mm in depth) at the center of the groove using a bone chisel with an outer diameter of 5.2 mm in minipigs. The columnar construct was carefully autografted into the osteochondral defect in one of the two defects (implanted defect), and nothing was implanted into the second defect (control). For obtaining the results of these studies, pigs were scanned by CT every 3 months following implantation and then euthanized at 24 or 48 weeks to confirm the osteochondral regeneration process by macroscopic and histological evaluation.

In CT images of MMPigs, the reduction in the subchondral radiolucent area of the implanted site became more dramatic at 2 or 3 months following surgery compared with that at the control site. A radiopaque area emerged from the boundary between the bone and the implant, and the area increased more steadily upward and inward for the implanted defect as time passed until 6 months after the surgery, compared with the control site. Thereafter, the radiopaque area of the implant gradually progressed and then filled the entire osteochondral defect at 12 months after surgery. In contrast, in the control site, a radiopaque area emerged in the shallow layer, but subchondral bone formation was not completed to any degree in the deep layer. Macroscopic examination at 6 months following implantation revealed that the surface of the implanted defect was covered with abundant cartilaginous white tissues, whereas cartilaginous tissue was scarce and the surface was depressed in the control site. Histopathological sections showed that thickened fibrocartilage had developed over the subchondral bone that was regenerating in the implanted site. The surface of the cartilage was smooth and the boundary with the surrounding normal cartilage was obscure in the implanted site. In comparison, the surface was collapsed and irregular in the control site. At 12 months following implantation, the surface was uniformly covered with abundant cartilaginous white tissues, and the boundary to the surrounding normal cartilage was unclear in the implanted site upon macroscopic examination. Histological examination indicated that the surface of the cartilage was smooth and the boundary with the surrounding normal cartilage was obscure, although small areas of endochondral ossification persisted at the center of the implanted site (Fig. 3c). Additionally, subchondral bone was symmetrically reconstructed in the implanted defect and was covered by a mixed matrix of hyaline cartilage and fibrocartilage (Fig. 3c). In the control site, although fibrocartilage had immediately covered the defect, the subchondral ossification was poor.

2.3 Rabbit Osteochondral Regeneration with Allogenic AT-MSCs

In this section, a study regarding a construct comprised of allogeneic AT-MSCs engrafted into osteochondral defects in rabbit models is presented [63]. Skeletally mature female Japanese white rabbits were used as the cell donor and defect recipients, which were divided into two groups according to treatment (Table 1). AT-MSCs were extracted from the AT of the interscapular fat pad of the donor female (Table 1). The harvested AT was washed, cut into small pieces, and digested in 0.12% type I collagenase, and the resulting solution was filtered and centrifuged. The pellet was resuspended in culture medium containing DMEM with 10% FBS and then plated onto a culture dishes and cultured for 1 week (Table 1). To generate AT-MSC spheroids, the cells were seeded into low-attachment 96 well plates at 5×10^4 cells/well (Table 1). Two days following incubation, the cells aggregated into a spheroid formation approximately 700 mm in diameter (Table 1). About 800 spheroids were added into a "mold" of 4.6 mm in diameter in culture medium (Table 1; Fig. 1a, b). These loaded spheroids adhered to one another and formed an AT-MSC construct after further culturing for several days (Fig. 1c–e). Implant surgery was performed using aseptic techniques and an osteochondral lesion (4.8 mm in diameter and 3 mm in depth) was created at the center of the trochlear groove using a drill. For the implanted group, the allogenic AT-MSC constructs were gently implanted into the defect (Fig. 3a, b). In the control group, the osteochondral defect was left empty. Rabbits were euthanized 4, 8, and 12 weeks following implantation to obtain the results of macroscopic and histological evaluation of healed tissues.

The microscopic findings of the osteochondral defects revealed that maturation of articular cartilage was noted at 4 weeks and increased gradually over time in the experimental group (Fig. 3c). In contrast, the surface of the created defect was covered with fibrous tissue, and healing tissue with large fissures was observed at 4 weeks after the operation in the control group. Therefore, the experimental group showed greater evidence of integration through cartilage-like tissue. In addition, inflammatory cells were not detected in the implanted defects. In comparison, fissures partially filled with fibrous tissue were observed in the control group.

2.4 Scaffold-Free 3D Constructs Produced Using the "Mold" Method

In this section, examples of histological examination for MSC constructs prepared by using the "mold" method were mentioned [60, 63] (Fig. 1e). Microscopic observations of the columnar constructs revealed that the spheroids agglutinated partially with each other within the construct. Moreover, numerous cell nuclei were confirmed both inside and around the spheroids without fragmentation or chromatin condensation, suggesting that the cells in the constructs were viable and did not undergo apoptosis. However, it was necessary to carefully grip the constructs using surgery tweezers because ECM such as type I collagen was present in only small amounts. The construct showed low expression of proteoglycan, which suggested that the cells in the constructs had not been differentiated into chondrocytes prior to the implantation. Based on these results, it was confirmed that AT-MSCs secreted type I collagen in order to form a self-generated scaffold, which migrated to fill in the interval spaces between each spheroid, and then formed a columnar construct prior to implantation. Consistent with this, the ultrastructural morphology of the constructs as assessed by scanning electron microscopy revealed that these constructs contained flattened cells with some amount of collagen fibers.

3 Osteochondral Regeneration with Scaffold-Free 3D Constructs Produced by the "Kenzan" Method

3.1 Swine Osteochondral Regeneration with Autologous AT-MSCs

In this section, a representative study is demonstrated that analyzed osteochondral regeneration following the autologous implantation of two swine AT-MSC constructs [71]. Skeletally mature male minipigs were used in this study, and autologous AT was aseptically excised under general anesthesia (Table 1). AT was immediately minced and digested in 0.1% collagenase, then the cell suspensions were filtered and centrifuged. The obtained pellet was suspended with culture medium consisting of DMEM containing 10% FBS and seeded into a culture flask (Table 1). Cells were incubated for 4–5 days and harvested at subconfluence as passage 0. Next, the harvested cells were resuspended in a special culture medium in which a xeno-free and serum-free MSC culture medium and a reduced serum (2%) MSC culture medium were mixed (Table 1). The cells were seeded into a culture flask and cultured until construct creation, with at least 3×10^7 autologous AT-MSCs being used to fabricate each construct (Fig. 1a). For this purpose, the AT-MSCs were resuspended in the special culture medium, and 1×10^4 cells/well were dispensed into low-attachment 96 well plates (Table 1). Following incubation for 24 h, the cells formed spheroids with a diameter of approximately 550 mm (Table 1). After the AT-MSC spheroids were prepared, a Bio-3D printer (Regenova®; Cyfuse Biomedical K.K., Tokyo, Japan) was used to assemble the AT-MSCs into scaffold-free tubular tissue constructs according to a 3D model predesigned on a computer system using a Bio-3D designer (B3D; Cyfuse Biomedical K.K., Tokyo, Japan). The bio-3D printer automatically skewers the AT-MSCs spheroids onto two types of "Kenzan," comprising needle grids of 9×9 (small) and 13×13 (large) with needle diameter of 0.17 mm and inter-needle interval of 0.4 mm (Table 1). The usable size of the small needle array was 3.4 mm$^2 \times$ 10 mm high and that of the large needle array was 5 mm$^2 \times$ 10 mm high (Table 1; Fig. 2a). In this system, the MSC spheroids were picked up separately from the 96 well plate by a robotically controlled fine suction 26 gauge nozzle.

Sequentially, the nozzle was inserted into the needle array to skewer the spheroid onto the "Kenzan" (Fig. 2b). To create two constructs according to the pre-designed tubular configuration, a total of 3,000 spheroids were prepared. After skewering the spheroids onto the "Kenzan," they were placed into the perfusion chamber and cultured in the special culture medium for 1 week to fuse the spheroids (Fig. 2c). The flow rate of the medium was 4–4.5 mL/min. After the spheroids were fused, the constructs were extracted from the "Kenzan," revealing a small tubular construct of 3.5 mm in diameter, 4 mm in height, and 1.5 mm in thickness and a large construct of 5 mm, 4 mm, and 1.5 mm, respectively (Fig. 2d, e). The implant surgery was performed under general anesthesia, and the articular cartilage and subchondral bone were holed to a depth of 4 mm at the center of the groove in both hind limbs using a bone chisel with an outer diameter of 5.2 mm. A large tubular construct was inserted into the defect site in the right hind limb, with a small construct inserted inside the large construct to complete the graft (Fig. 4a, b). Alternatively, the left hind limb was left untreated as a control. For obtaining the results of this study, pigs were scanned by CT and magnetic resonance (MR) imaging every 3 months following implantation and then euthanized at 24 weeks to confirm the osteochondral regeneration process by macroscopic and histological evaluation.

A radiopaque area was found at the boundary between the bone and the graft, which increased steadily upward and inward in the defect sites implanted with constructs as compared with the control sites. In both implanted and control defect sites, the area was reduced at 3 and at 6 months after surgery, as compared with the area immediately after surgery. In addition, the area at 3 months was lower in the implanted defects than that in the control defects. MR images of the implanted defect sites showed restoration of the articular cartilage, with signal patterns similar to that of the surrounding normal cartilage. Nevertheless, new bone formation under the cartilaginous tissue was incomplete. The distinction between the superficial and deep cartilage layers, as shown by the high- and the low-signal intensity layers, respectively, was nearly restored, and AT predominantly occupied the subchondral area in the implanted sites. Macroscopic examination revealed that the surfaces of the osteochondral defects were better restored in the implanted sites than those of the control

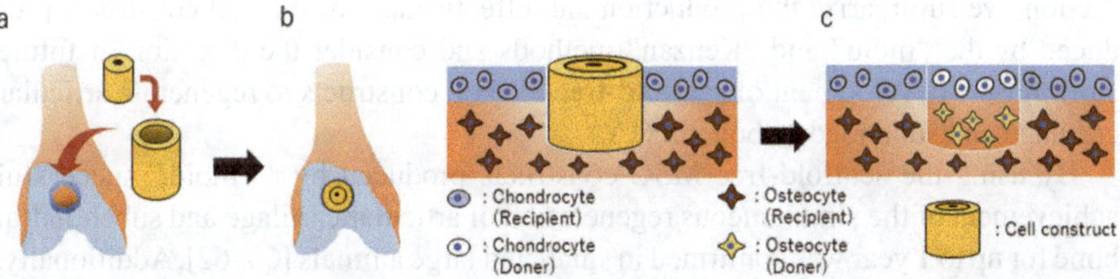

Fig. 4 Scaffold-free cell constructs produced by using the "Kenzan" method for osteochondral regeneration in vivo. (**a**) Scaffold-free small cell construct was inserted into scaffold-free large cell construct for the implantation. (**b**) Scaffold-free cell constructs were implanted into an osteochondral defect. (**c**) Mesenchymal stem cells in the constructs differentiated to chondrocytes and osteocytes, followed by regeneration of articular cartilage in the surface layer and formation of subchondral bone in the deep layer at the implanted site

sites. Histopathology showed smooth hyaline cartilage, and subchondral bone formation was also noted in the implanted sites (Fig. 4c). In contrast, a deeply recessed surface and a lower rate of bone formation were observed in the control defects.

3.2 Scaffold-Free 3D Constructs Produced Using the "Kenzan" Method

In this section, an example is indicated that the histology of AT-MSC constructs prepared using bio-3D printing using the Kenzan method was evaluated [71] (Fig. 2e). Histology of the small tubular constructs showed that the spheroids were fused with each other within the construct and sufficient type I collagen was produced in the constructs and almost filled in the spaces between the spheroids. Consequently, the constructs exhibited moderate hardness so that they could be gripped easily with surgery tweezers. The construct exhibited low expression of proteoglycan, whereas the majority of the cells in the construct were negative for TdT-mediated dUTP nick end labeling (TUNEL) staining with very few positive cells around the needle holes. Based on these results, it was confirmed that AT-MSCs secreted abundant type I collagen in order to form a self-generated scaffold, which formed a rigid columnar construct.

4 Conclusion

In this chapter, four types of studies were introduced, which demonstrated osteochondral regeneration using scaffold-free cell constructs consisting of (1) autologous rabbit BM-MSC spheroids generated by using a "mold" [60], (2) autologous swine AT-MSC spheroids created by using a "mold" [61, 62], (3) allogenic rabbit AT-MSC spheroids produced by using a "mold" [63], and (4) autologous swine AT-MSC spheroids fabricated by using the "Kenzan" method [71]. In the following section, we summarize the production and effectiveness of the cell constructs produced by the "mold" and "Kenzan" methods and consider the direction of future work for the development of scaffold-free 3D cell constructs to regenerate articular cartilage and subchondral bone in OA.

By using the scaffold-free MSC constructs produced by a "mold," successful achievement of the simultaneous regeneration of articular cartilage and subchondral bone for up to 1 year was confirmed in small and large animals [60–62]. Additionally, the scaffold-free allogeneic MSC constructs implanted into the osteochondral defects survived, adhered to the defect, and regenerated articular cartilage and subchondral bone in vivo [63]. Although allogeneic cells have the limitation that they will be foreign to the immune system of the recipient, it has been suggested that AT-MSCs and their secretions afford the potential to induce immunologic tolerance

[72] and that allogeneic MSC transplantation does not enhance the hyper-response of T cells against donor antigens [69]. Furthermore, implantation of an artificial scaffold-free autologous MSC construct fabricated using a Bio-3D printer with the "Kenzan" method into an osteochondral defect can aid in the regeneration of articular cartilage and subchondral bone in a large animal [71].

For the reconstruction of articular cartilage and subchondral bone in the above studies, spheroids were made only of MSCs because it is difficult to expand differentiated cells and as yet impossible to produce the target tissues in vitro [71]. Therefore, stem cells were used for producing pluripotent cell constructs to differentiate according to the surrounding environment following implantation [71]. However, although BM-MSCs and AT-MSCs were employed for the purpose of those studies, other stem cells with higher chondrogenic differentiation abilities such as SM-MSCs have been reported [34]. In addition, spheres consisting only of induced pluripotent stem (iPS) cell-derived chondrocytes have been reported to contribute to cartilage regeneration in minipigs in only 1 month [73]. Therefore, other types of MSCs and differentiated component cells from iPS cells can be considered as potential cell sources for construct fabrication.

Although the medium used to produce the cell constructs generated by using a "mold" constituted a basic culture medium for MSCs, the mixed medium used for the construct made by using the "Kenzan" method comprised a unique medium that has not been reported for use in any other studies [71]. As it was found that the cell constructs produced in this experiment were filled with connective tissue consisting of abundant type I collagen, the mixed medium appeared to contribute to inducing the MSCs to produce type I collagen and improve the strength and elasticity of the construct [71]. Moreover, the constructs were intended to be generated without using any differentiation agents, and the cells in the constructs were undifferentiated when the constructs were implanted into the osteochondral defect, supporting this conjecture [71]. However, as the components of the utilized media were not specified, their relationship with the construct properties should be examined in detail in future studies. In this respect, it should also be confirmed that the AT-MSCs in the construct maintained their undifferentiated status immediately prior to implantation.

The columnar construct in the "mold" was restrictively exposed to the medium at only the upper and bottom areas, whereas the tubular construct fixed with the "Kenzan" could be more widely exposed at the outer and inner surfaces [60–63, 71]. Furthermore, when the spheroids are arranged onto the "Kenzan" in a constant volume by 3D printing, the packing ratio (density) would be lower than that calculated when the spheroids are packed in the same volume of a "mold" at random (Figs. 1 and 2). Therefore, more medium could perfuse among the spheroids in a tubular compared with a columnar construct. However, if the outer and inner diameters of large and small tubular constructs and the diameter of osteochondral defects are mismatched, gaps could be formed between the defect and the large tubular construct, between the large and small tubular constructs, or at the center of the small construct and consequently might induce differences in tissue regeneration [71]. Therefore, it may be necessary to develop a method for more efficiently supplying culture medium to the columnar constructs.

Histology revealed that the construct was negative for type II collagen in ECM [60, 63, 71]. This appears not to be necessary for cartilage repair, although the production of type II collagen would be increased according to the chondrogenesis in the construct because most of the cells were viable (i.e., negative for TUNEL staining) except for cells close to the needle holes. All results of the studies in which the constructs produced using the "mold" and "Kenzan" methods were implanted into osteochondral defects showed that the deep layer of the construct was differentiated into the bone and the surface layer was differentiated into cartilage, depending on the surrounding environment at the implanted defect site [60–63, 71] (Figs. 3 and 4). Cell tracking analysis also revealed that chondrocytes and osteocytes in the regenerated tissues were derived dominantly from the implanted MSCs within the construct rather than the migrated cells from the surrounding tissues [60, 63]. Thus, the MSCs were induced to differentiate into chondrocytes in vivo and later secreted type II collagen. Similarly, it is possible that proteoglycan was produced by the chondrocytes derived from AT-MSCs during the cartilage regeneration process following implantation. Thus, the constructs presented were considered to acquire the in vivo characteristics and mechanical strength of cartilage via the ECM that was produced during the maturation process into osteocartilage following adhesion to the recipient tissue (Figs. 3 and 4). However, the morphological (i.e., detailed observation by electron microscopy) and functional (non-destructive test to evaluate the physical properties) evaluation of the repaired/regenerated tissues have not yet been reported. In addition, the underlying molecular mechanisms of the methods using scaffold-free AT-MSC constructs, in which osteochondral defects were regenerated while maintaining the border between articular cartilage and subchondral bone, remain unclear.

Nevertheless, these observations suggest that in the future, it will be possible to produce cell constructs exhibiting optimal functions as implants for osteochondral reconstruction by optimizing the cell types and medium for construct fabrication. Moreover, it is also expected that studies using larger osteochondral defect models will be performed and the mechanism of articular cartilage regeneration and subchondral bone formation will be evaluated by assessing the regeneration process in detail over time. And then, I hope that the cell construct fabricated by the "Kenzan" method will be applied clinically as a promising treatment for OA.

References

1. Ono N, Ono W, Nagasawa T et al (2014) A subset of chondrogenic cells provides early mesenchymal progenitors in growing bones. Nat Cell Biol 16:1157–1167
2. Flowers SA, Zieba A, Örnros J et al (2017) Lubricin binds cartilage proteins, cartilage oligomeric matrix protein, fibronectin and collagen II at the cartilage surface. Sci Rep 7:13149
3. Young B, Lowe JS, Steavens A et al (2006) Skeletal tissues. In: Young B, Lowe JS, Steavens A, Heath JW (eds) Wheater's functional histology: a text and color atlas, 5th edn. Elsevier, Oxford, pp 186–206
4. Heir S, Nerhus TK, Røtterud JH et al (2010) Focal cartilage defects in the knee impair quality of life as much as severe osteoarthritis: a comparison of knee injury and osteoarthritis outcome score in 4 patient categories scheduled for knee surgery. Am J Sports Med 38:231–237

5. Minzlaff P, Feucht MJ, Saier T et al (2016) Can young and active patients participate in sports after osteochondral autologous transfer combined with valgus high tibial osteotomy? Knee Surg Sports Traumatol Arthrosc 24:1594–1600
6. Muraki S, Oka H, Akune T et al (2009) Prevalence of radiographic knee osteoarthritis and its association with knee pain in the elderly of Japanese population-based cohorts: the ROAD study. Osteoarthr Cartil 17:1137–1143
7. Gelber AC, Hochberg MC, Mead LA et al (2000) Joint injury in young adults and risk for subsequent knee and hip osteoarthritis. Ann Intern Med 133:321–328
8. Fernandes JC, Martel-Pelletier J, Pelletier JP (2002) The role of cytokines in osteoarthritis pathophysiology. Biorheology 39:237–246
9. Ding C, Cicuttini F, Scott F et al (2006) Natural history of knee cartilage defects and factors affecting change. Arch Intern Med 166:651–658
10. Muraki S, Akune T, Oka H et al (2009) Association of occupational activity with radiographic knee osteoarthritis and lumbar spondylosis in elderly patients of population-based cohorts: a large-scale population-based study. Arthritis Rheum 61:779–786
11. Mankin HJ (1982) The response of articular cartilage to mechanical injury. J Bone Joint Surg Am 64:460–466
12. Lane JG, Massie JB, Ball ST (2004) Follow-up of osteochondral plug transfers in a goat model: a 6-month study. Am J Sports Med 32:1440–1450
13. Szerb I, Hangody L, Duska Z et al (2005) Mosaicplasty: long-term follow-up. Bull Hosp Jt Dis 63:54–62
14. Bentley G, Biant LC, Carrington RW et al (2003) A prospective, randomised comparison of autologous chondrocyte implantation versus mosaicplasty for osteochondral defects in the knee. J Bone Joint Surg Br 85:223–230
15. Matsusue Y, Yamamuro T, Hama H et al (1993) Arthroscopic multiple osteochondral transplantation to the chondral defect in the knee associated with anterior cruciate ligament disruption. Arthroscopy 9:318–321
16. Steadman JR, Briggs KK, Rodrigo JJ et al (2003) Outcomes of microfracture for traumatic chondral defects of the knee: average 11-year follow-up. Arthroscopy 19:477–484
17. Brittberg M, Lindahl A, Nilsson A et al (1994) Treatment of deep cartilage defects in the knee with autologous chondrocyte transplantation. N Engl J Med 331:889–895
18. Huntley JS, Bush PG, McBirnie JM et al (2005) Chondrocyte death associated with human femoral osteochondral harvest as performed for mosaicplasty. J Bone Joint Surg Am 87:351–360
19. Kreuz P, Steinwachs M, Erggelet C et al (2006) Results after microfracture of full-thickness chondral defects in different compartments in the knee. Osteoarthr Cartil 14:1119–1125
20. Fujisato T, Sajiki T, Liu Q et al (1996) Effect of basic fibroblast growth factor on cartilage regeneration in chondrocyte-seeded collagen sponge scaffold. Biomaterials 17:155–162
21. Funayama A, Niki Y, Matsumoto H et al (2008) Repair of full-thickness articular cartilage defects using injectable type II collagen gel embedded with cultured chondrocytes in a rabbit model. J Orthop Sci 13:225–232
22. Diaz-Romero J, Gaillard JP, Grogan SP et al (2005) Immunophenotypic analysis of human articular chondrocytes: changes in surface markers associated with cell expansion in monolayer culture. J Cell Physiol 202:731–742
23. Duan L, Liang Y, Ma B et al (2017) DNA methylation profiling in chondrocyte dedifferentiation in vitro. J Cell Physiol 232:1708–1716
24. Knutsen G, Engebretsen L, Ludvigsen TC et al (2004) Autologous chondrocyte implantation compared with microfracture in the knee. A randomized trial. J Bone Joint Surg Am 86:455–464
25. Tatebe M, Nakamura R, Kagami H et al (2005) Differentiation of transplanted mesenchymal stem cells in a large osteochondral defect in rabbit. Cytotherapy 7:520–530
26. Pittenger MF, Mackay AM, Beck SC et al (1999) Multilineage potential of adult human mesenchymal stem cells. Science 284:143–147

27. Zuk PA, Zhu M, Ashjian P et al (2002) Human adipose tissue is a source of multipotent stem cells. Mol Biol Cell 13:4279–4295
28. Nimura A, Muneta T, Koga H et al (2008) Increased proliferation of human synovial mesenchymal stem cells with autologous human serum: comparisons with bone marrow mesenchymal stem cells and with fetal bovine serum. Arthritis Rheum 58:501–510
29. Im GI, Kim DY, Shin JH et al (2001) Repair of cartilage defect in the rabbit with cultured mesenchymal stem cells from bone marrow. J Bone Joint Surg Br 83:289–294
30. Nam H, Karunanithi P, Loo WC et al (2013) The effects of staged intra-articular injection of cultured autologous mesenchymal stromal cells on the repair of damaged cartilage: a pilot study in caprine model. Arthritis Res Ther 15:R129
31. Zhou G, Liu W, Cui L et al (2006) Repair of porcine articular osteochondral defects in non-weightbearing areas with autologous bone marrow stromal cells. Tissue Eng 12:3209–3221
32. Zuk PA, Zhu M, Mizuno H et al (2001) Multilineage cells from human adipose tissue: implications for cell-based therapies. Tissue Eng 7:211–228
33. Mochizuki T, Muneta T, Sakaguchi Y et al (2006) Higher chondrogenic potential of fibrous synovium- and adipose synovium-derived cells compared with subcutaneous fat-derived cells: distinguishing properties of mesenchymal stem cells in humans. Arthritis Rheum 54:843–853
34. Nakamura T, Sekiya I, Muneta T et al (2012) Arthroscopic, histological and MRI analyses of cartilage repair after a minimally invasive method of transplantation of allogeneic synovial mesenchymal stromal cells into cartilage defects in pigs. Cytotherapy 14:327–338
35. Fraser JK, Hicok KC, Shanahan R et al (2014) The Celution system: automated processing of adipose-derived regenerative cells in a functionally closed system. Adv Wound Care (New Rochelle) 3:38–45
36. Gentile P, Scioli MG, Orlandi A et al (2015) Breast reconstruction with enhanced stromal vascular fraction fat grafting: what is the best method? Plast Reconstr Surg Glob Open 3:e406
37. Koga H, Muneta T, Ju YJ et al (2007) Synovial stem cells are regionally specified according to local microenvironments after implantation for cartilage regeneration. Stem Cells 25:689–696
38. Koga H, Shimaya M, Muneta T et al (2008) Local adherent technique for transplanting mesenchymal stem cells as a potential treatment of cartilage defect. Arthritis Res Ther 10:R84
39. Suzuki S, Muneta T, Tsuji K et al (2012) Properties and usefulness of aggregates of synovial mesenchymal stem cells as a source for cartilage regeneration. Arthritis Res Ther 14:R136
40. Wakitani S, Kimura T, Hirooka A et al (1989) Repair of rabbit articular surfaces with allograft chondrocytes embedded in collagen gel. J Bone Joint Surg (Br) 71:74–80
41. Wakitani S, Goto T, Pineda SJ et al (1994) Mesenchymal cell-based repair of large, full-thickness defects of articular cartilage. J Bone Joint Surg Am 76:579–592
42. Lu Z, Doulabi BZ, Huang C et al (2010) Collagen type II enhances chondrogenesis in adipose tissue-derived stem cells by affecting cell shape. Tissue Eng Part A 16:81–90
43. Yoon IS, Chung CW, Sung JH, Cho HJ, Kim JS, Shim WS, Shim CK, Chung SJ, Kim DD (2011) Proliferation and chondrogenic differentiation of human adipose-derived mesenchymal stem cells in porous hyaluronic acid scaffold. J Biosci Bioeng 112:402–408
44. Chen WC, Yao CL, Wei YH et al (2011) Evaluating osteochondral defect repair potential of autologous rabbit bone marrow cells on type II collagen scaffold. Cytotechnology 63:13–23
45. Unterman SA, Gibson M, Lee JH et al (2012) Hyaluronic acid-binding scaffold for articular cartilage repair. Tissue Eng Part A 18:2497–2506
46. Arrigoni E, De-Girolamo L, Di-Giancamillo A et al (2013) Adipose-derived stem cells and rabbit bone regeneration: histomorphometric, immunohistochemical and mechanical characterization. J Orthop Sci 18:331–339
47. Park CW, Rhee YS, Park SH et al (2010) In vitro/in vivo evaluation of NCDS-micro-fabricated biodegradable implant. Arch Pharm Res 33:427–432
48. Robinson D, Efrat M, Mendes DG et al (1993) Implants composed of carbon fiber mesh and bone-marrow-derived, chondrocyte-enriched cultures for joint surface reconstruction. Bull Hosp Jt Dis 53:75–82

49. Nakayama K (2013) In vitro biofabrication of tissues and organs. In: Forgacs G, Sun W (eds) Biofabrication: micro- and nano-fabrication, printing, patterning and assemblies, 1st edn. Elsevier, Oxford, pp 1–21

50. Kierszenbaum AL (2002) Basic tissues and integrated cell biology. In: Kierszenbaum AL, Tres LL (eds) Histology and cell biology: an introduction to pathology, 3rd edn. Elsevier, Oxford, pp 113–145

51. Anderson JM, Rodriguez A, Chang DT (2008) Foreign body reaction to biomaterials. Semin Immunol 20:86–100

52. Badylak SF, Gilbert TW (2008) Immune response to biologic scaffold materials. Semin Immunol 20:109–116

53. Patrascu JM, Freymann U, Kaps C et al (2010) Repair of a post-traumatic cartilage defect with a cell-free polymer-based cartilage implant: a follow-up at two years by MRI and histological review. J Bone Joint Surg Br 92:1160–1163

54. O'Brien FJ (2011) Biomaterials & scaffolds for tissue engineering. Mater Today 14:88–95

55. Kaneshiro N, Sato M, Ishihara M et al (2006) Bioengineered chondrocyte sheets may be potentially useful for the treatment of partial thickness defects of articular cartilage. Biochem Biophys Res Commun 349:723–731

56. Ando W, Tateishi K, Hart DA et al (2007) Cartilage repair using an in vitro generated scaffold-free tissue-engineered construct derived from porcine synovial mesenchymal stem cells. Biomaterials 28:5462–5470

57. Cheuk YC, Wong MWN, Lee KM et al (2011) Use of allogeneic scaffold-free chondrocyte pellet in repair of osteochondral defect in a rabbit model. J Orthop Res 29:1343–1350

58. Dean DM, Napolitano AP, Youssef J et al (2007) The directed self-assembly of microtissues with prescribed microscale geometries. FASEB J 21:4005–4012

59. Rago AP, Dean DM, Morgan JR (2009) Controlling cell position in complex heterotypic 3D microtissues by tissue fusion. Biotechnol Bioeng 102:1231–1241

60. Ishihara K, Nakayama K, Akieda S et al (2014) Simultaneous regeneration of full-thickness cartilage and subchondral bone defects in vivo using a three-dimensional scaffold-free autologous construct derived from high-density bone marrow-derived mesenchymal stem cells. J Orthop Surg Res 9:98

61. Murata D, Tokunaga S, Tamura T et al (2015) A preliminary study of osteochondral regeneration using a scaffold-free three-dimensional construct of porcine adipose tissue-derived mesenchymal stem cells. J Orthop Surg Res 10:35

62. Murata D, Akieda S, Misumi K et al (2017) Osteochondral regeneration with a scaffold-free three-dimensional construct of adipose tissue-derived mesenchymal stromal cells in pigs. Tissue Eng Regen Med 15:101–113

63. Oshima T, Nakase J, Toratani T et al (2019) A scaffold-free allogeneic construct from adipose-derived stem cells regenerates an osteochondral defect in a rabbit model. Arthroscopy 35:583–593

64. Kawaguchi H, Miyoshi N, Miura N et al (2011) Microminipig, a non-rodent experimental animal optimized for life science research: novel atherosclerosis model induced by high fat and cholesterol diet. J Pharmacol Sci 115:115–121

65. Takeishi K, Horiuchi M, Kawaguchi H et al (2012) Acupuncture improves sleep conditions of minipigs representing diurnal animals through an anatomically similar point to the acupoint (GV20) effective for humans. Evid Based Complement Alternat Med 2012:472982

66. Jurgens WJ, Kroeze RJ, Zandieh-Doulabi B et al (2013) One-step surgical procedure for the treatment of osteochondral defects with adipose-derived stem cells in a caprine knee defect: a pilot study. Biores Open Access 2:315–325

67. De Girolamo L, Niada S, Arrigoni E et al (2015) Repair of osteochondral defects in the minipig model by OPF hydrogel loaded with adipose-derived mesenchymal stem cells. Regen Med 10:135–151

68. Pojda Z (2015) Adipose-derived stem cells for therapeutic applications. In: Bhattacharya N, Stubblefield PG (eds) Regenerative medicine, 1st edn. Springer, London, pp 77–89
69. Mahmoud EE, Tanaka Y, Kamei N et al (2018) Monitoring immune response after allogeneic transplantation of mesenchymal stem cells for osteochondral repair. J Tissue Eng Regen Med 12:275–286
70. Itoh M, Nakayama K, Noguchi R et al (2013) Scaffold-free tubular tissues created by a bio-3D printer undergo remodeling and endothelialization when implanted in rat aortae. PLoS One 10:e0136681
71. Yamasaki A, Kunitomi Y, Murata D et al (2019) Osteochondral regeneration using constructs of mesenchymal stem cells made by bio three-dimensional printing in mini-pigs. J Orthop Res 37:1398–1408
72. Lee SM, Lee SC, Kim SJ (2014) Contribution of human adipose tissue-derived stem cells and the secretome to the skin allograft survival in mice. J Surg Res 188:280–289
73. Yamashita A, Morioka M, Yahara Y et al (2015) Generation of scaffoldless hyaline cartilaginous tissue from human iPSCs. Stem Cell Rep 4:404–418

Scaffold-Free Biofabrication of Liver

Yusuke Yanagi, Toshiharu Matsuura, and Tomoaki Taguchi

Abstract The liver performs multiple functions that are essential for life. The development of a functional human liver tissue model showing a similar response to that in the human body for investigations of liver disease and medical treatment has been challenging. Many studies have described in vitro human liver tissue models based on two-dimensional cultures of primary hepatocytes. However, those models lost their function rapidly. Efforts to engineer a three-dimensional liver tissue model are currently underway. Regenerative therapy using engineered liver tissue has been proposed as a promising alternative to orthotopic liver transplantation for patients presenting with liver failure and liver-based inherited metabolic diseases. To increase its potential clinical utility, the next step will be to make the human liver model large enough to reflect the liver function in the human body. The key to constructing larger tissue specimens is to avoid creating an ischemic environment ex vivo. In this respect, the "Kenzan" method has advantages through its elaborate geometry and immediate execution of culture circulation. We successfully fabricated human liver-like tissue exhibiting self-tissue organization and engraftment on the liver of a rat using the "Kenzan" method. Our efforts may provide new insight into cell-based therapy and facilitate the shift from cell- to bio-tissue-based therapy.

Keywords End-stage liver disease · Liver transplantation · Hepatocyte transplantation · Liver tissue engineering · Liver bud · Liver organoid · 3D bioprinter

Y. Yanagi (✉) · T. Matsuura · T. Taguchi
Department of Pediatric Surgery, Reproductive and Developmental Medicine, Kyushu University Graduate School of Medical Sciences, Fukuoka, Japan
e-mail: y-yanag@pedsurg.med.kyushu-u.ac.jp

© Springer Nature Switzerland AG 2021
K. Nakayama (ed.), *Kenzan Method for Scaffold-Free Biofabrication*,
https://doi.org/10.1007/978-3-030-58688-1_6

Abbreviations

NPCs   Non-parenchymal cells
ECM   Extracellular matrix
OLT   Orthotopic liver transplantation
2D   Two-dimensional
3D   Three-dimensional
iPS   Induced pluripotent stem
ES   Embryonic stem
hHep   Human mature hepatocyte
HUVEC   Human umbilical vein endothelial cell
hMSC   Human mesenchymal stem cell
LBS   Liver-bud-like spheroid

1   Introduction

The liver is the heaviest internal organ in the human body, typically accounting for 2–3% of the body weight. The liver is a unique organ, receiving blood supply through the portal vein as well as from the intestine, spleen, and pancreas. The classic microscopic unit of the liver is the hepatic lobule, with its central vein, portal triads (portal vein, hepatic artery, and bile duct), and intervening sinusoids [1].

The liver is a complex organ involving multiple cells working together. It is constructed of two major types of cell: parenchymal cells and non-parenchymal cells (NPCs). The majority of the cells are parenchymal hepatocytes, accounting for approximately 70–85% of the liver volume. Hepatocytes are the main functional cells of the liver and the primary metabolic component [2]. The hepatic lobule is highly organized with hepatocytes at the interface between endothelial sinusoids and bile canaliculi.

NPCs, by contrast, include cholangiocytes, sinusoidal endothelial cells, hepatic stellate cells, and Kupffer cells, among others, and not only provide support for the hepatocytes but also contribute to homeostatic or inflammatory responses through their own functions. The bile duct is composed of cholangiocytes. Sinusoidal endothelial cells line the hepatocytes and play a significant role in the transport of molecules from the circulating blood to the hepatocytes. Hepatic stellate cells are found in the space of Disse and involved in the formation of the extracellular matrix (ECM). Kupffer cells exhibit inflammatory and immune response [3].

The liver has a complex function, being involved in metabolism, excretion, and body defense [4]. The liver is also a center for coordinating energy homeostasis. Key roles of hepatic metabolism include protein synthesis coupled with amino acid metabolism, carbohydrate metabolism, lipid metabolism, and biotransformation. The liver also plays an important role in the metabolism of vitamins, trace elements such as copper, and peptide hormones [4, 5]. Bile acid is synthesized in the liver and excreted with numerous exogenous and endogenous substances through the biliary tract. Furthermore, the liver defends the body against toxic chemicals and bacteria entering through the portal vein, playing crucial roles all around.

2 Human Liver Tissue Model

A functional human liver tissue model showing responses similar to those observed in the in vivo human environment is needed for in vitro liver experiments. Many studies have described in vitro human liver tissue models based on two-dimensional (2D) cultures of human primary hepatocytes. However, those models typically lost their function rapidly, as it is difficult for primary hepatocytes to proliferate and maintain their function for a long period in vitro after isolation [6, 7].

The three-dimensional (3D) structure of the liver reportedly helps maintain its function. The success of 3D reaggregate and spheroid culture systems in vitro suggests that cells may require cell–cell contact and a proper 3D cytoarchitecture similar to that found in vivo to achieve the optimum function. Spheroid culture has been able to maintain a greater liver-specific function than monolayer culture for a longer period owing to the formation of cell junctions, cytoskeletal dynamics, and intracellular trafficking [8–10]. Furthermore, sandwich cultures of primary hepatocytes have shown that the maintenance of hepatocyte polarity requires the interplay of ECM interactions in addition to cell adhesion and formation of cell junctions [11, 12]. It has also become clear that communication and support from various NPCs helps hepatocytes retain their function [13, 14]. Cell aggregates containing multiple types of cells have thus attracted considerable attention as living materials or cellular building blocks for fabricating 3D liver tissue models [15, 16].

Drug discovery research is the most feasible and important field for adopting human liver models in vitro. A human liver tissue model that sustains the drug metabolism function can predict drug hepatotoxicity and metabolite production in humans [17]. Several models exhibiting the drug metabolism function have been reported, including a micropatterned model [18, 19], a spheroid model [20, 21], and a model constructed with 3D bioprinting [22, 23]. Furthermore, if a human liver tissue model could be generated from patient-derived cells, such as primary hepatocytes or induced pluripotent stem cells (iPS cells), individualized safety, pharmacokinetics, and identification of viable drugs could be achieved.

3 Clinical Applications of Liver Tissue Models
   for Regenerative Medicine

3.1 Liver Disorder

The liver performs multiple functions that are essential for life, as mentioned in the previous section. Liver disease can be caused by a variety of factors that damage the liver, such as genetics, infection, immune system, cancer, drugs, and alcohol use. Disease development, regardless of the cause, can result in the loss of liver functions, which causes significant disorder to the body.

Approximately 500,000 people have liver cirrhosis, and 170,000 patients die annually because of cirrhosis in Japan. In the United States, approximately

30 million people have liver disorders, which are responsible for 30,000 deaths annually [24, 25]. The inability of hepatocyte may get into failure of maintaining metabolism, synthesizing proteins like albumin and coagulation factors, and conjugating and secreting bilirubin. Liver failure is characterized by portal hypertension, coagulopathy, and hepatic encephalopathy [26, 27] and is a life-threatening condition eventually requiring orthotopic liver transplantation (OLT). OLT has become the gold standard treatment for patients presenting with acute or chronic liver failure and liver-based inherited metabolic diseases when conservative treatment fails to restore or maintain a sufficient liver function. Although OLT still has some issues with perioperative mortality and posttransplant morbidities, the outcomes are generally excellent and significantly prolong the patient survival. However, donor organ shortages have limited its benefit, as organ demand exceeds supply by manyfold [28]. Organ shortages along with the invasiveness of surgery and poor mortality/morbidity rates associated with OLT have prompted the exploration of alternative treatments to OLT.

3.2 Hepatocyte Transplantation

Since the indispensable function of the liver is not a mechanical function but a metabolic one, repopulation of hepatocytes may enable the regeneration of the liver function. Hepatocyte transplantation has been proposed as an alternative to OLT over the past several decades [29]. This treatment is less invasive and less radical than OLT because cells can be injected into the liver through the portal vein while the recipient's native liver is left in place, with the cells subsequently integrating into the host liver. A number of clinical trials of hepatocyte transplantation have shown its therapeutic effect in cases of acute liver failure, chronic liver failure, and inherited liver disease [30, 31]. However, while a complete cure was achieved in some cases, the results were limited to indicate its important role as a bridging treatment to OLT through avoiding neurological symptom and emergent liver transplant in most cases [32].

Low-level engraftment of the transplanted cells and lack of long-term efficacy are the major obstacles to successful replacement of the deficient liver function by hepatocyte transplantation [33, 34]. Problems associated with hepatocyte transplantation are as follows: the number of cells deemed transplantable via endovascular transplantation has been severely restricted due to portal hypertension and embolism [35], and the direct injection of hepatocytes into blood vessels resulted in the activation of tissue factors accelerating clot formation and evoking an immune reaction [34]. In addition, procuring a sufficient number of hepatocytes for transplantation has proven difficult. Hepatocytes for cell transplantation have traditionally been harvested from discarded donor livers or discarded liver tissue generated through hepatectomy [36, 37]. While robust ex vivo expansion of hepatocytes has drawn recent attention [38, 39], it has not yet been universally achieved. The shortage of donor livers as a cell source is also a major obstacle in this field. Novel cell sources, such as somatic stem cells [40–42], embryonic stem (ES) cells, and iPS cells, have recently shown promise for generating functional hepatocyte-like cells [43, 44].

Remaining issues that must be overcome in cell-based therapies include producing and transplanting a sufficient number of cells, avoiding vascular complications, accelerating the engraftment of cells, and maintaining the function of the transplanted cells.

3.3 Liver Tissue Model Transplant

Recently, tissue engineering technologies have been investigated as an alternative approach to cell-based therapy. Several reports on engineered liver tissues, such as a hepatic tissue sheets and a liver tissue device, have described the engraftment and therapeutic effect of this approach in vivo [[45], [46]]. The most notable study on liver tissue engineering concerned the fabrication of a "liver bud," which was an aggregate of iPS-derived endodermal cells combined with umbilical vein endothelial cells and mesenchymal stem cells [47]. The mixed cells of multiple types aggregated and self-organized into 3D spheroid structures termed a "liver bud." This multiple-type cell interaction and 3D morphology promoted hepatocyte differentiation of human iPS-derived endodermal cells, and these in vitro-generated liver buds showed similar DNA profiles to actual liver buds, demonstrating outstanding therapeutic potential in mice [48].

To increase the potential clinical utility of these engineered liver tissue models, the next step will be to make the human liver model large enough to reflect the liver function in the human body. Several approaches are available for constructing a scalable 3D cellular structure using cell aggregates or spheroids as building blocks. Indeed, many studies have focused on seeding the spheroid into three-dimensional scaffolds, such as decellularized liver and a scaffold constructed by artificial materials [49]. However, the usage of exogenous materials in scaffold is associated with several issues, including infection, immune reactions, and the degradation of materials and biofilms [50]. Therefore, a scaffold-free tissue engineering approach will likely show greater clinical safety.

4 Liver Tissue Engineering Using the "Kenzan" Method

We have highlighted the interesting general capacity of cells to self-organize and form structures with similar histological properties, such as those seen in vivo [51]. Dissociated cells are able to assemble through the autonomous cell-to-cell direct adhesion process. This cell–cell adhesion phenomenon in tissue reconstitution has been recognized for over 100 years [52] as a critical process for tissue morphogenesis and organ development, carried out with cell assemblage, sorting, and migration by cadherin-mediated adhesion process [53]. This phenomenon has also been adapted in recent studies on regenerative medicine.

We tried to fabricate a scalable liver tissue using liver-bud-like spheroids (LBSs) containing human mature hepatocytes (hHeps), human umbilical vein endothelial cells (HUVECs), and human mesenchymal stem cells (hMSCs) as building blocks [54]. Various technologies for assembling cellular aggregates to fabricate 3D tissues have been reported [55, 56]. However, those previous technologies have had some disadvantages associated with the materials used and conditions of tissue culture. Since the liver is a solid tissue with high cellularity, the engineered liver tissue may be fabricated in a fashion wherein the majority of cells are arranged close to each other without leaving any void spaces or ECM. The cell–cell interactions at the time of construction lead to the more rapid achievement of tissue-like morphology and functionality [49].

We selected the "Kenzan" method because it requires only cells as materials (a scaffold-free method) and is capable of rapidly and automatically fabricating a transplantable scaffold-free 3D tissue in an arbitrary shape [57, 58].

We first established the appropriate culture conditions for generating a large number of LBSs of uniform shape in order to be able to perform 3D printing stably. The microscopic observation of the spheroid formation showed that the HUVECs were mainly localized at the central regions. hHeps assembled slowly by interacting with MSCs and were located at the circumference, while MSCs were evenly distributed (Fig. 2a).

Ensuring sufficient nutrients and oxygen was an important issue for the construction of the 3D tissue [59, 60]. The LBSs fabricated in our study had a diameter of approximately 500 µm. If the hepatocyte spheroids were to be used in vitro, the viable diameter would be limited to approximately 100–150 µm due to the depletion of oxygen [61–63]. In this respect, using a mixture of multiple cells offered several benefits. We noted that hHeps were located at the circumference, while HUVECs and MSCs were located at the core. This distribution was expected to alleviate the hypoxic condition of the hHeps. The combination of HUVECs and MSCs promoted vascular formation through the release of vascular endothelial growth factor (VEGF) and hepatocyte growth factor (HGF) under hypoxic conditions [64, 65]. Furthermore, we fabricated a tube-shaped structure with spacing between each of the LBSs to supply fresh medium all over the structure and avoid creating an ischemic environment ex vivo. This spacing was so delicate that LBSs were in contact with each other in the diagonal direction and had as much space in the vertical direction as possible (Fig. 1a, b). The ability to create this microscale space is one of the major advantages of the "Kenzan" method, enabling the fabrication of elaborate geometry in 3D tissue.

After forming LBSs, we bioprinted them to fabricate a tube-shaped tissue. The LBSs on the "Kenzan" were fused and then removed from the needle array. After continuous culture, a scalable 3D structure was obtained (Fig. 1c).

Hematoxylin–eosin staining of the structure showed that the hHeps were viable, even in the deep area of the tissue (Fig. 2b). ECM was detected between LBSs that were considered to have been produced by the cell itself. The cell–matrix interaction is essential for tissue formation and the maintenance of the cell function besides cell–cell interaction. Our approach enabled the fostering of this important interaction without using any exogenous matrix, which ensures the clinical safety.

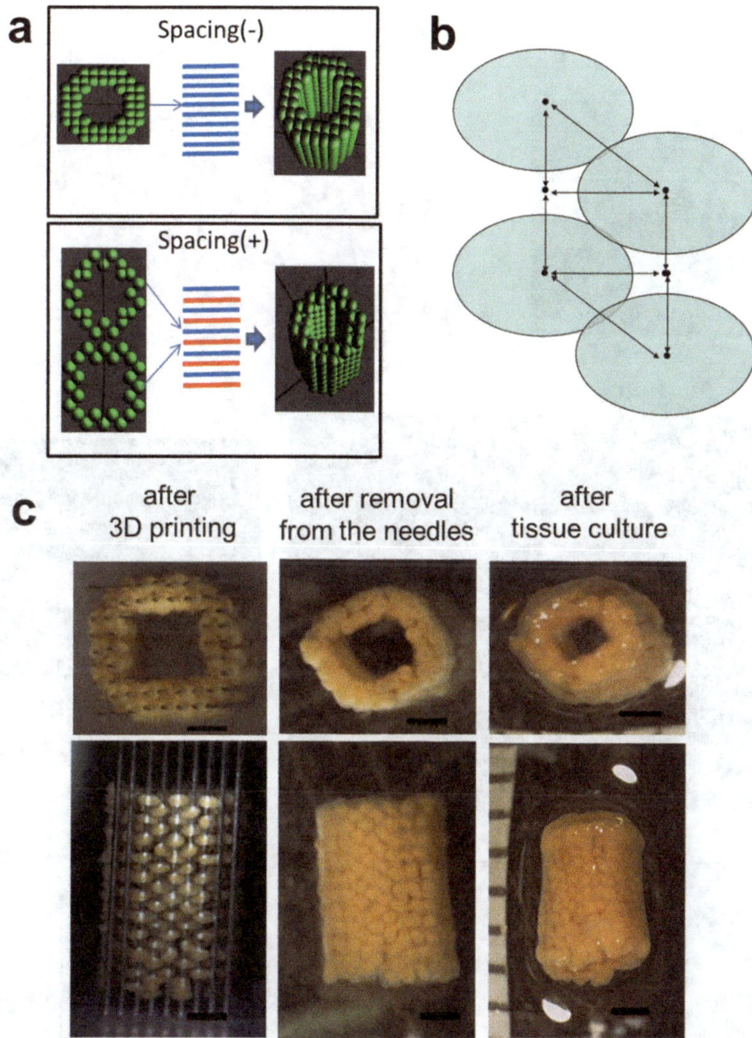

Fig. 1 The construction of a scalable liver-like tissue using the "Kenzan" method. (**a**) Comparative designs of the tube-shaped construct with and without spacing. (**b**) The figure of the space between spheroids. (**c**) Biofabricated tube-shaped tissue and morphological changes during 3D tissue construction. Scale bar, 200 μm. (Reproduced from [54])

Immunofluorescence microscopy revealed the production of albumin by hHeps and the formation of a reticular network by the HUVECs (Fig. 2c, d). After 1 month of culture circulation, the CK19-positive cells that were transdifferentiated from primary hepatocytes were observed in the 3D structure, and some of them had formed a duct-like structure (Fig. 2e). The expression of several hepatic genes showed a significant increase after 3D culture with the medium circulation. The level of albumin secreted by the tissue with circulation was significantly increased and remained at a high level for 1 month (Fig. 2f). Based on these findings, we concluded that the

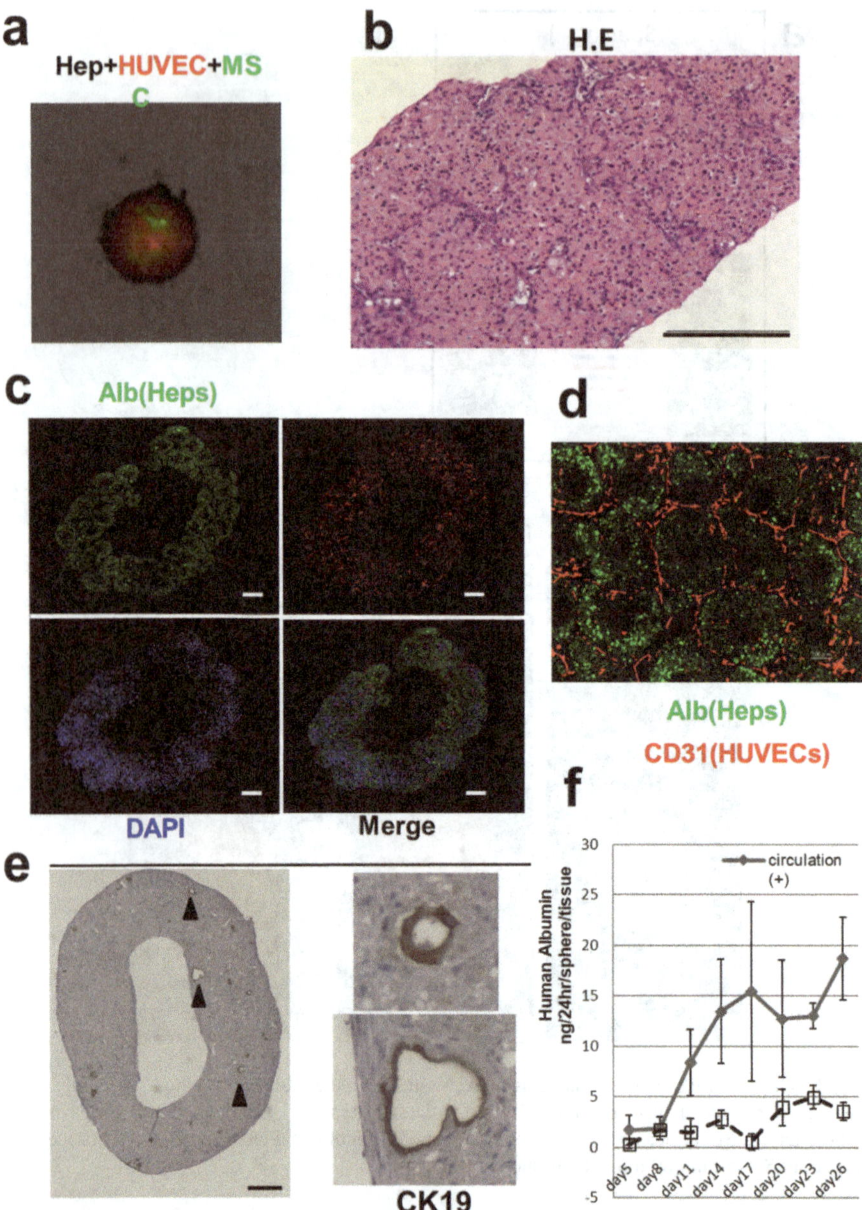

Fig. 2 The ex vivo-fabricated liver-like tissue. (**a**) The microscopic observation of liver-bud-like spheroids (hHeps, black; HUVECs, red; and MSCs, green). (**b**) An HE-stained cross section on the seventh day in culture. The hepatocytes were viable, even in the center of the tissue, the thickness of which was approximately 600 μm. Scale bars, 100 μm. (**c,d**) Fluorescence images of the liver-like tissue. hHeps were stained with antihuman albumin antibodies (green), and HUVECs were stained with antihuman CD31 antibodies (red). The production of albumin by hHeps and the formation of reticular endothelial networks inside the liver-like tissue were verified.

3D structure was growing into a liver-like tissue through ex vivo tissue self-organization [54].

Our hierarchical scaling process from cells to spheroids to tissue was achieved through this fundamental process of cell–cell adhesion and 3D spatial arrangement. While the underlying molecular mechanisms are unclear, cells were able to self-assemble and self-sort to guide tissue formation. Since the fusion of spheroids took several days, as shown in our study, the fixation of the shape with the needle-array system during the period of spheroid fusion is one of the major advantages of the "Kenzan" method. Some methods of constructing scaffold-free structures using spheroids used an outer frame to fix the spheroids [66]. However, this use of an outer frame to maintain the geometry of the assembled spheroids likely makes it difficult to construct a stable structure with culture circulation.

Finally, we confirmed the transplantability of the biofabricated liver-like tissue. We directly connected the biofabricated liver-like tissue to the liver of a nude rat and subsequently detected the production of human albumin by hHeps [54].

5 Future Expectation

As we described in the introduction, the liver comprises sophisticated structures that incorporate specialized hepatic components, such as bile canaliculi and sinusoids. The production of canalicular and sinusoidal networks that are functionally connected will require greater effort. Furthermore, stem cell-derived cells should be used to overcome the cell source shortage.

The ideal patients to receive treatment via liver tissue transplantation are those with inherited metabolic liver disease, especially children [67]. Inherited metabolic liver diseases are caused by a deficiency in a single hepatic enzyme or protein, thus leading to hepatic and/or extrahepatic disease. However, the remaining liver functions are normal. We learned through our studies that compensation of just 5–15% of the missing protein amount through hepatocyte transplantation can correct the metabolic disease. It is therefore theoretically unnecessary to replace the entire organ, as replacement of simply a small portion of the liver mass is effective in achieving the desired outcome [68].

We hope that this new approach to liver tissue transplantation will provide new insight into cell-based therapy for liver diseases, thus facilitating the shift from cell- to tissue-based applications. In the future, full liver replacement will hopefully be achieved through similar means.

Fig. 2 (continued)   Scale bars, 200 μm. (**e**) Immunostaining of the liver-like tissue on the 24th day of culture. CK19-positive cells and a duct-like morphology were observed. The arrowheads indicate the CK19-positive cells with a duct-like morphology. Scale bars, 200 μm. (**f**) The comparison of the secretion of albumin from the biofabricated tissue with/without culture circulation, as measured by an ELISA. Circulation (+), n = 4; circulation (−), n = 3. The results represent the mean ± SD. *$p < 0.05$. (Reproduced from [54])

Acknowledgments The authors thank Mr. Brian Quinn for help in the preparation of the English version of this article.

References

1. Kiernan F (1833) The anatomy and physiology of the liver. Philos Trans R Soc Lond Biol 123:711–770
2. Si-Tayeb K et al (2010) Organogenesis and development of the liver. Dev Cell 18:175–189
3. Bale SS et al (2016) Isolation and co-culture of rat parenchymal and non-parenchymal liver cells to evaluate cellular interactions and response. Sci Rep 6:25329
4. Corless JK, Middleton HM 3rd (1983) Normal liver function. A basis for understanding hepatic disease. Arch Intern Med 143:2291–2294
5. Brauer RW (1963) Liver circulation and function. Physiol Rev 43:115–213
6. Jasmund I et al (2007) The influence of medium composition and matrix on long-term cultivation of primary porcine and human hepatocytes. Biomol Eng 24:59–69
7. Berry MN et al (1992) Techniques for pharmacological and toxicological studies with isolated hepatocyte suspensions. Life Sci 51:1–16
8. Landry J et al (1985) Spheroidal aggregate culture of rat liver cells: histotypic reorganization, biomatrix deposition, and maintenance of functional Activities. J Cell Biol 101:914–923
9. Goulet F et al (1988) Cellular interactions promote tissue-specific function, biomatrix deposition and junctional communication of primary cultured hepatocytes. Hepatology 8:1010–1018
10. Chang TT, Hughes M (2014) Molecular mechanisms underlying the enhanced functions of three-dimensional hepatocyte aggregates. Biomaterials 35:2162–2171
11. Gissen P, Arias IM (2015) Structural and functional hepatocyte polarity andl iver disease. J Hepatol 63:1023–1037
12. Zeigerer A et al (2017) Functional properties of hepatocytes in vitro are correlated with cell polarity maintenance. Exp Cell Res 350:242–252
13. Clement B et al (1984) Long-term co-cultures of adult human hepatocytes with rat liver epithelial cells: modulation of albumin secretion and accumulation of extracellular material. Hepatology 4:373–380
14. Lu HF et al (2005) Three-dimensional co-culture of rat hepatocyte spheroids and NIH/3T3 fibroblasts enhances hepatocyte functional maintenance. Acta Biomater 1:399–410
15. Guven S et al (2015) Multiscale assembly for tissue engineering and regenerative medicine. Trends Biotechnol 33:269–279
16. Mironov V et al (2009) Organ printing: tissue spheroids as building blocks. Biomaterials 30:2164–2174
17. Lin C, Khetani SR (2016) Advances in engineered liver models for investigating drug-induced liver injury. Biomed Res Int:1829148
18. Khetani AR, Bhatia SN (2008) Microscale culture of human liver cells for drug development. Nat Biotechnol 26:120–126
19. Wang WW et al (2010) Assessment of a micropatterned hepatocyte coculture system to generate major human excretory and circulating drug metabolites. Drug Metab Dispos 38:1900–1905
20. Friedrich J et al (2009) Spheroid-based drug screen: considerations and practical approach. Nat Protocol 4:309–324
21. Bell CC et al (2016) Characterization of primary human hepatocyte spheroids as a model system for drug-induced liver injury, liver function and disease. Sci Rep 6:25187
22. Nguyen DG et al (2016) Bioprinted 3D primary liver tissues allow assessment of organ-level response to clinical drug induced toxicity in vitro. PLoS One 11:e0158674
23. Kizawa H et al (2017) Scaffold-free 3D bio-printed human liver tissue stably maintains metabolic functions useful for drug discovery. Biochem Biophys Rep 10:186–191

24. Alqahtani SA (2012) Update in liver transplantation. Curr Opin Gastroenterol 28:230–238
25. Mokdad AA et al (2012) Liver cirrhosis mortality in 187 countries between 1980 and 2010: a systematic analysis. BMC Med 12:145
26. Gines P et al (2004) Management of cirrhosis and ascites. N Engl J Med 350:1646–1654
27. Schuppan D, Afdhal NH (2008) Liver cirrhosis. Lancet 371:838–851
28. Zarrinpar A, Busuttil RW (2013) Liver transplantation: past, present and future. Nat Rev Gastroenterol Hepatol 10:434–440
29. Forbes SJ et al (2015) Cell therapy for liver disease: from liver transplantation to cell factory. J Hepatol 62:S157–S169
30. Matas AJ et al (1976) Hepatocellular transplantation for metabolic deficiencies: decrease of plasma bilirubin in Gunn rats. Science 192:892–894
31. Fox IJ et al (1988) Treatment of the Crigler-Najjar syndrome type I with hepatocyte transplantation. N Engl J Med 338:1422–1426
32. Enosawa S et al (2014) Hepatocyte transplantation using a living donor reduced graft in a baby with ornithine transcarbamylase deficiency: a novel source of hepatocytes. Liver Transpl 20:391–393
33. Tiffany N et al (2018) Hepatocyte transplantation: past efforts, current technology, and future expansion of therapeutic potential. J Surg Res 226:48–55
34. Puppi J et al (2012) Improving the techniques for human hepatocyte transplantation: report from a consensus meeting in London. Cell Transplant 21:1–10
35. Ponder KP et al (1991) Mouse hepatocytes migrate to liver parenchyma and function indefinitely after intrasplenic transplantation. Proc Natl Acad Sci U S A 88:1217–1221
36. Mitry RR et al (2004) One liver, three recipients: segment IV from split-liver procedures as a source of hepatocytes for cell transplantation. Transplantation 77:1614–1646
37. Enosawa S (2017) Isolation of GMP grade human hepatocytes from remnant liver tissue of living donor liver. *Transplantation*. Methods Mol Biol 1506:231–245
38. Zhang K et al (2018) In vitro expansion of primary human hepatocytes with efficient liver repopulation capacity. Cell Stem Cell 23:806–819
39. Hu H et al (2018) Long-term expansion of functional mouse and human hepatocytes as 3D organoids. Cell 175:1591–1606
40. Yamaza T et al (2015) In vivo hepatogenic capacity and therapeutic potential of stem cells from human exfoliated deciduous teeth in liver fibrosis in mice. Stem Cell Res Ther 6:171
41. Hu X et al (2016) Direct induction of hepatocyte-like cells from immortalized human bone marrow mesenchymal stem cells by overexpression of HNF4α. Biochem Biophys Res Commun 478:791–797
42. Chen G (2015) Adipose-derived stem cell-based treatment for acute liver failure. Stem Cell Res Ther 6:40
43. Takahasi K (2007) Induction of pluripotent stem cells from adult human fibroblasts by defined factors. Cell 131:861–872
44. Sekiya S, Suzuki A (2011) Direct conversion of mouse fibroblasts to hepatocyte-like cells by defined factors. Nature 475:390–393
45. Ohashi K et al (2007) Engineering functional two- and three-dimensional liver systems in vivo using hepatic tissue sheets. Nat Med 13:880–885
46. Gutierrez AS et al (2010) Engineering of a hepatic organoid to develop liver assist devices. Cell Transplant 19:815–822
47. Takebe T et al (2013) Vascularized and functional human liver from an iPSC-derived organ bud transplant. Nature 499:481–484
48. Takebe T et al (2014) Generation of a vascularized and functional human liver from an iPSC-derived organ bud transplant. Nat Protocol 9:396–409
49. Collin de l'Hortet A et al (2016) Liver-regenerative transplantation: regrow and reset. Am J Transplant 16:1688–1696
50. Nakayama K (2013) Biofabrication. In: Forgacs J (ed) Micro- and nano-fabrication, printing, patterning, and assemblies. Elsevier, Amsterdam, pp 1–16

51. Sasai Y (2013) Cytosystems dynamics in self-organization of tissue architecture. Nature 493:318–326
52. Wilson H (1907) On some phenomena of coalescence and regeneration in sponges. J Exp Zool 5:245–258
53. Takeichi M (2011) Self-organization of animal tissues: cadherin-mediated processes. Dev Cell 21:24–26
54. Yanagi Y et al (2017) *In vivo* and *ex vivo* methods of growing a liver bud through tissue connection. Sci Rep 7:14085
55. Norotte C et al (2009) Scaffold-free vascular tissue engineering using bioprinting. Biomaterials 30:5910–5917
56. Török E et al (2011) Primary human hepatocytes on biodegradable poly(l-Lactic acid) matrices: a promising model for improving transplantation efficiency with tissue engineering. Liver Transpl 17:104–114
57. Itou M (2015) Scaffold-free tubular tissues created by a bio-3D printer undergo remodeling and endothelialization when implanted in rat aortae. PLoS One 10:e0145971
58. Yurie H et al (2017) The efficacy of a scaffold-free bio 3D conduit developed from human fibroblasts on peripheral nerve regeneration in a rat sciatic nerve model. PLoS One 10:e0171448
59. Sekine H et al (2013) In vitro fabrication of functional three-dimensional tissues with perfusable blood vessels. Nat Commun 4:1399
60. Pang Y et al (2012) Liver tissue engineering based on aggregate assembly: efficient formation of endothelialized rat hepatocyte aggregates and their immobilization with biodegradable fibres. Biofabrication 4:045004
61. Glicklis R et al (2004) Modeling mass transfer in hepatocyte spheroids via cell viability, spheroid size, and hepatocellular functions. Biotechnol Bioeng 86:672–680
62. Curcio E et al (2007) Mass transfer and metabolic reactions in hepatocyte spheroids cultured in rotating wall gas-permeable membrane system. Biomaterials 28:5487–5497
63. Anada T (2012) An oxygen-permeable spheroid culture system for the prevention of central hypoxia and necrosis of spheroids. Biomaterials 33:8430–8441
64. Huang CC et al (2013) Hypoxia-induced therapeutic neovascularization in a mouse model of an ischemic limb using cell aggregates composed of HUVECs and cbMSCs. Biomaterials 34:9441–9450
65. Chen DY et al (2013) Intramuscular delivery of 3D aggregates of HUVECs and cbMSCs for cellular cardiomyoplasty in rats with myocardial infarction. J Control Release 172:419–425
66. Matsunaga YT et al (2011) Molding cell beads for rapid construction of macroscopic 3D tissue architecture. Adv Mater 23:H90–H94
67. Taguchi T et al (2019) Regenerative medicine using stem cells from human exfoliated deciduous teeth (SHED): a promising new treatment in pediatric surgery. Surg Today 49:316–322
68. Johns C et al (2012) Hepatocyte transplantation for inherited metabolic diseases of the liver. J Intern Med 272:201–223

Artificial Trachea: Past, Present, and Future

Keitaro Matsumoto and Takeshi Nagayasu

Abstract The trachea and bronchus play important roles in the human body, providing a pathway for air to reach the lungs. Various diseases can arise in the airway, including cancer, stenosis, and tracheobronchomalacia. Surgery represents the main form of treatment for these diseases. However, resecting over 50% of the tracheal length is currently unfeasible because of the need for remnant trachea. To overcome this limitation and improve curability rates, alternatives are needed. Many investigators in both research fields and clinical practice have reported on artificial trachea. Autologous tissue, synthetic materials, and allograft have all been used as artificial trachea. However, practical artificial trachea for use in the clinical field remains lacking. Issues with artificial trachea that need to be overcome include infection, rigidity, the need for immunosuppressants, and in vivo growth for pediatric cases. Scaffold-free artificial trachea made using the patient's own cells could resolve some of these issues. Artificial trachea created using the "Regenova" bio-3D printing system has been successfully transplanted into rats to replace native trachea, achieving survival for more than 1 year without immunosuppressants. Many issues of trachea made by foreign materials have thus been solved in rats. However, many issues remain to be addressed before clinical application.

Keywords Trachea replacement · Artificial trachea · Surgery · Regeneration

K. Matsumoto (✉) · T. Nagayasu
Department of Surgical Oncology, Nagasaki University Graduate School of Biomedical Sciences, Nagasaki, Japan

Medical-Engineering Hybrid Professional Development Center, Nagasaki University Graduate School of Biomedical Sciences, Nagasaki, Japan
e-mail: kmatsumo@nagasaki-u.ac.jp

© Springer Nature Switzerland AG 2021
K. Nakayama (ed.), *Kenzan Method for Scaffold-Free Biofabrication*,
https://doi.org/10.1007/978-3-030-58688-1_7

1 Introduction

The trachea and bronchus are parts of the airway extending from the larynx to the lung. The airway is an organ providing a pathway for air to move in and out of the lungs. This is a critical organ in the human body, providing a lumen for the exchange of oxygen and carbon dioxide in the lungs. Trouble arising in the airway can restrict respiration and thus limit air exchange in the lungs. This can easily lead to life-threatening situations. In addition, the airway plays a role in trapping foreign materials before they reach the lungs, because the airway is always exposed to outside air and airborne materials, including dust, bacteria, viruses, and fungi. The trachea and bronchus are therefore not only air ducts but also a form of barrier to the respiratory system and human body. This multiplicity of function leads to the trachea and bronchus having complicated structures. Regeneration of the airway represents a major goal for clinical use and thus represents a challenging and thrilling theme for researchers around the world. In this chapter, we discuss the topics of airway replacement and bio-3D-printed trachea.

2 Anatomy and Function of the Airway

The respiratory system comprises several parts, including the respiratory tract, central nervous system, pulmonary artery, pulmonary vein, chest wall, and accessory muscles of respiration. The respiratory tract can be divided into several segments: the conducting airways including the trachea and proximal bronchioles, the respiratory bronchioles, and the alveoli. The trachea is an organ with a duct extending from the larynx to the lungs allowing the air flow for the exchange of oxygen and carbon dioxide between air and blood in the lung. In humans, the trachea develops from endoderm and mesoderm [1].

The airway branches repeatedly (approximately 14 times) to form conduits for air to reach several distinct pulmonary segments. The trachea bifurcates at the carina into the right and left main bronchi. In adults, the mean external diameter of the trachea is 2.3 and 1.8 cm in coronal and sagittal dimensions, respectively. The trachea comprises incomplete cartilaginous tracheal rings and smooth muscle, with deficiency in the cartilage at the posterior of the tracheal rings adjacent to the esophagus. The role of these cartilaginous rings is to keep the trachea patent. The tracheal rings are connected by smooth muscle. The adult trachea has a length of around 11.8 cm with a normal range of 10–13 cm and is composed of approximately 18–22 D-shaped anterior cartilaginous rings [2] (Fig. 1).

The blood supply to the trachea divides into cervical and thoracic parts. The tracheoesophageal branches from the inferior thyroid arteries and thyrocervical arteries supply blood to the cervical trachea, while the thoracic trachea and carina receive blood flow via the bronchial arteries from the aorta. In general, the superior, middle, and inferior bronchial arteries supply blood flow to the trachea and carina.

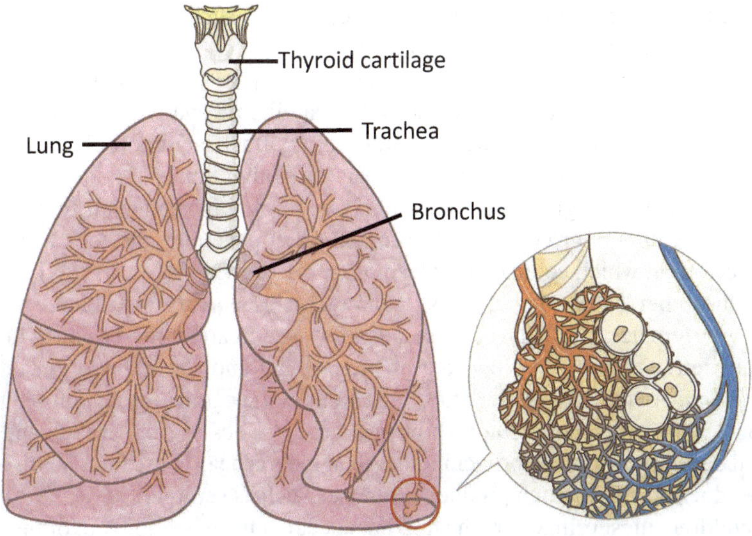

Fig. 1 Anatomy of the airway and lung

A number of superior bronchial arteries arise from intercostal bronchial and internal thoracic arterial trunks [2].

The right and left recurrent laryngeal nerves are branches of the vagus nerve and function to innervate the true vocal cords. They enter the larynx between the thyroid and cricoid cartilages.

The adult human airway is covered by a continuous epithelial layer comprising less than 1% of the total respiratory epithelial surface [3–5]. The airway epithelium is stratified in the large airways, while columnar and cuboidal epithelium is evident in the small airways. The cell types are ciliated, columnar, undifferentiated, secretory, and basal cells. The luminal mucosa of the trachea is lined by ciliated pseudostratified columnar epithelium containing mucus-producing goblet cells. The mucosa also harbors ducts that connect mucous glands in the submucosa to the surface of the tracheal lumen. The surface mucus and cilia act to trap and expel particulate matter and microorganisms that enter the airway [2].

3 Diseases

Several types of malignant and benign tracheal disease can occur, such as tracheal tumors, tracheal stenosis, and tracheobronchomalacia.

Primary tracheal tumors comprise approximately 0.1–0.4% of malignant diseases, with around 2.6 new cases per million people per year [6–9], but only 8% of these tumors occur in children [10, 11]. This means the relative incidence of tracheal tumors compared to bronchial tumors is under 2% [12]. Primary tracheal tumors can develop from a wide variety of cell types, such as epithelium, salivary glands, and mesenchymal structures of the trachea [13, 14]. Around 90% of primary tumors are malignant in adults [15], compared to only 10–30% in children [10, 11]. Regarding pathological types, squamous cell carcinoma and adenoid cystic carcinoma account for about two-thirds of adult primary tracheal tumors. The remaining one-third includes various types of malignant or benign tumor [13, 14]. Squamous cell carcinoma is strongly associated with cigarette smoking and tends to develop in males. These tumors grow relatively fast and often penetrate into the airway wall with ulceration, which could cause bleeding in the trachea.

On the other hand, adenoid cystic carcinoma is equally distributed between males and females, partially because adenoid cystic carcinoma is not associated with cigarette smoking. A slow rate of progression and low penetration into the tracheal wall are other characteristics of adenoid cystic carcinoma. Carcinoid tumor is another type of tracheal tumor, showing characteristics of slow growth and preferential development in the bronchi rather than the trachea. These tumors arise from neuroendocrine cells, which produce hormones such as serotonin.

In children, most primary tracheal tumors occur at the posterior wall of the cervical trachea, mostly between the neonatal period and 14 years old [11]. Benign tumors include papillomas, chondromas, and hemangiomas. Papillomas are the most common type of benign tracheal tumor in children. These are caused by human papillomavirus and could transform into squamous cell carcinoma.

Tracheal stenosis and tracheobronchomalacia are other pathologies that are more frequent in children.

Tracheal stenosis, which includes subglottic stenosis, involves a narrowing or constriction of the trachea. Many cases of tracheal stenosis could be caused by scar tissue due to prolonged intubation. Other causes include external injury, autoimmune disorders, and infections. This pathology can also develop as a side effect of radiotherapy used to treat tumors.

Tracheobronchomalacia is a clinical condition with airway collapse during breathing. Common symptoms include coughing, wheezing, shortness of breath, difficulty clearing phlegm, and respiratory infections. Causes of tracheobronchomalacia vary depending on the case. Most cases of primary tracheobronchomalacia are caused by underlying genetic conditions that weaken the airway walls, such as mucopolysaccharidoses, Ehlers-Danlos syndrome, and a variety of chromosomal abnormalities. Primary tracheobronchomalacia can be idiopathic, associated with prematurity or certain birth defects [16–18].

Acquired tracheobronchomalacia is generally caused by degeneration of the cartilages supporting the airways. In adults, the cause usually cannot be identified [19]. Many adult patients with tracheobronchomalacia show common respiratory conditions such as asthma, chronic bronchitis, and emphysema [19]. Acquired tracheobronchomalacia may be associated with inflammatory conditions, exposure to toxins, enlargement of structures near the airway, and complications from medical procedures [16–19].

4 Conventional Treatment

The treatments for tracheal diseases depend on the type and condition of tracheal disease. Types of treatment include surgery, radiation, chemotherapy, and endoscopic treatments such as stenting, ablation, ballooning, and forceps extraction.

For benign and low-grade malignant tracheal tumors, single treatment by surgery can achieve complete cure. Surgery could therefore provide these patients with long-term survival and pathological confirmation of complete tumor removal and relieve airway obstruction permanently [15, 20].

The decision on whether to use surgery or radiation for a tracheal tumor depends on several factors, including the condition of the patient, and tumor factors such as histology, location and occupation area. Basically, resectability should be estimated before any other treatment. However, if life-threatening airway obstruction is found, bronchoscopic treatment with or without stenting should be chosen. Depending on the patient, resection can be performed after bronchoscopic treatment [21]. Chemotherapy and radiotherapy are other options for patients who do not meet the indications for surgery [15, 20].

Surgical resection is the most important treatment for tracheal disease. Various surgical treatments can cure primary tumors of the airway from the larynx to the carina, including laryngectomy, tracheal resection, and carinal resection with or without lungs. Endoscopic resection or laser or electrocautery fulguration for benign diseases is usually used in children. These various surgical procedures have been described in detail previously [22–28]. Preservation of the blood supply to the lateral segment is important, so the surrounding tissues should be handled gently, and anastomosis should be performed with utmost care. Another surgical point is the balance between oncological radicality and anastomotic tension. Tumor-bearing margins can be accepted in some radiotherapy-sensitive cases. To reduce anastomotic tension during surgery, mobilization of the trachea is of central importance in surgical procedures. About half of the total trachea can be removed by anatomical mobilization in adults with primary reconstruction. In children, about one-third of the trachea should be resectable. Therefore, for reconstruction after longer tracheal resection, a high-quality surgical technique is required. Even if multiple mobilization of the trachea is performed, the length of resectable trachea is limited. This limitation means that if alternatives are identified, curability rates should increase.

Extensive dissection of lymph nodes may not be recommendable, because this is likely to decrease the blood supply to the remaining trachea. However, no consensus has been made on this issue.

5 Regeneration and Replacement of Trachea: From Past to Present

Tracheal reconstruction and replacement are complicated and vital surgical procedures for several indications, as described above. Direct end-to-end anastomosis is considered insufficient after primary tracheal resection when the length of trachea resected is greater than 50% in adults or 30% in children. Tracheal replacement with an alternative is thus necessary to achieve perfect airway repair. The basics and principles of tracheal replacement were described in 1950 [29] and performed in 1979 [30]. However, even in the present era, tracheal replacement remains a challenging procedure because of the surrounding vital organs.

The ideal substitute for the airway should show the following characteristics: lateral rigidity and stiffness; longitudinal flexibility; airtight lumen; no need for immunosuppression; reliable, feasible, and producible technique; and biocompatible-like integration with adjacent tissues and healing. The presence of these characteristics could prevent chronic inflammation, granulation tissue, infection, and erosion [31]. The substitute should not involve the use of growth factors, which are basically contraindicated in patients with cancer. Further, in children, the tracheal replacement should increase in size with age in order to adapt to the patient's natural process of anatomical growth [32]. In addition, epithelial resurfacing would be ideal, because the replacement trachea should be able to play the roles of avoiding stenosis or late buckling and avoiding bacterial colonization and accumulation of secretions [33, 34].

In general, research attempts at tracheal replacement and regeneration have included the following: (1) wide variety of foreign materials with numerous technical modifications to avoid complications of implantation; (2) implantation of nonviable tissues, including fixed trachea; (3) adaptation and transfer of autogenous tissues with or without foreign-material scaffolding as patches or tubes; (4) tissue engineering of required components such as cartilage; and/or (5) transplantation of allografts with and without immunosuppressive therapy, preservation, and vascularization procedures. Varying degrees of success have been announced episodically over the decades in each of these categories, but this means that to date, no promising replacement methods have been applied in clinical practice over the long term in any safe and practicable manner [35].

Many techniques and substitutes have been used over the decades to create the ideal tracheal replacements. We describe and review the history of this area and topics leading to future clinical practice.

5.1 Autologous Tissue

Autologous tissue represents a development of classical autologous organ reconstruction. Two types of autologous tissue reconstruction are applied: pedicled flap and free flap. Pedicled flaps include greater omentum flap [36], intercostal muscle flap [37], pectoralis major muscle flap [38], and sternocleidomastoid muscle flap. These vascularized tissues offer the advantages of faster healing, greater reliability, and better stability over time. A combination of these grafts with a rigid structure provides a high-strength product to maintain an open lumen, such as trachea with cartilage rings. For example, reconstructions of the trachea have been reported using left radial forearm flap with biodegradable mesh rings or radial forearm free flap with combined PolyMax mesh (Synthes, Paoli, PA, USA) and Hemashield vascular graft (Boston Scientific, Natick, MA, USA) [39, 40].

Regarding free flaps, the largest clinical experience has been accumulated using free fascio-cutaneous flap from the forearm with cartilage struts [41, 42]. A key advantage of this method is the lack of risks of acute or chronic rejection. However, limitations to this method include the lack of spontaneous mucociliary clearance within the lumen and the risk of cartilage fracture among elderly patients [43]. This procedure was performed for 16 patients and reported a 5-year survival rate of 64.8%. Some complications were reported, such as chronic severe respiratory insufficiency with requiring stenting or permanent tracheostomy.

5.2 Synthetic Materials

Synthetic materials are, of course, non-biocompatible. Such materials would therefore increase the risk of granulation, infection, and erosion of adjacent organs [44]. Inflammation due to events such as infection may lead to adhesion or fistula with surrounding tissues or vessels such as the aorta or other great vessels, increasing the risk of life-threatening bleeding into the lungs [45, 46]. Synthetic materials have been used for more than a decade to maintain lumen rigidity. They constitute an interface in the respiratory tract, which could occur inflammatory granulomas that would in turn obstruct the lumen. Furthermore, these materials can become colonized with microorganisms, causing purulent debris and halitosis [47]. Prosthetic tracheal reconstructions have thus shown only limited success because of problems such as local infection, anastomotic leakage, dislocation, granuloma formation, and stenosis or rupture of adjacent vessels [46, 48–53].

5.3 Prosthesis

Some solid prostheses in clinical use have been reported. Neville et al. used a silicone rubber prosthesis and non-terminal Dacron ring [54]. In that trial, morbidity was high, with suture-line granulomas developing in 29% and graft dehiscence in one patient. Long-term outcomes were not able to be evaluated because of a lack of follow-up data. Toomes et al. [45] reported high postoperative mortality using a solid prosthesis. Based on these outcomes, solid prostheses appear difficult to use as a viable option for extensive tracheal replacement.

Porous prostheses may be more biocompatible than solid prostheses, because connective and epithelial tissues are able to grow into the prosthesis.

Polypropylene mesh, as a representative example, has been used in many investigations [55, 56]. In animals, uncoated porous polypropylene prostheses have been relatively successful for cervical tracheal replacement. In contrast, in clinical studies, tracheal reconstruction with a polypropylene mesh prosthesis has shown high morbidity and mortality rates, with patients showing lethal complications like fistula between the trachea and innominate artery, dehiscence, and stenosis of anastomosis [56]. As another material, a double ring of titanium was used in tracheal reconstruction. In some patients, this approach was feasible and not accompanied by any swallowing problems [57, 58]. However, this approach is limited to the proximal part of the trachea associated with the larynx, and cannot be considered for extended tracheal replacement.

On the other hand, collagen-conjugated Marlex mesh (pore size, 260 μm; C.R. Bard, Billerica, MA, USA) has shown feasible outcomes [59]. Liu et al. used porous-type tracheal prostheses reinforced with a continuous polypropylene spiral and sealed by collagen sponge from porcine skin for segments of cervical trachea. They concluded that this prosthesis is feasible because of the lack of need for omental wrapping and silicone stenting as ancillary measures. This approach could be indicated as a possible clinical application for cervical tracheal replacement.

5.4 Allograft

Allografts are compatible for organ replacement, as for organ transplantation. Aorta and trachea are candidates for replacement of the trachea. In general, immunosuppression would be necessary for transplantation, but decellularization of these organs allows immunosuppression after surgery to be avoided. On the other hand, lack of specific vascularization and rigidity are the key issues for replacement. Tracheal allograft offers a good candidate for tracheal replacement, but for circumferential grafts, the indications should remain limited.

5.5 Aortic Allografts

In animal models, aortic autografts and fresh or cryopreserved allografts could work as tracheal or bronchial substitutes with regeneration of cartilage and epithelium from the native airways, even though data are lacking regarding cartilage regeneration in autografted aorta [60, 61].

In one clinical study, fresh, cryopreserved, descending thoracic aorta from cadaveric donors was used [62]. For graft revascularization, the allograft was wrapped in a pectoralis major muscle flap with the attending thoraco-acromial blood supply. The morbidity rate was high, with events such as tracheoesophageal fistula, anterior spinal cord ischemia, multiple stent migration in the graft, and fatal massive hemoptysis from a fistula with a contiguous artery.

5.6   Tracheal Transplantation

In tracheal allograft transplantation, of course, immunological problems are critical in addition to technical difficulties and blood supply from surrounding tissues. Rejection could result in tracheal stenosis, necrosis, and other negative effects. In addition, of course, under immunosuppression, malignant tumor recurrence can result if the tracheal replacement is performed for patients with a history of malignant tumor. After some negative data in the early phase [30, 63, 64], Delaere et al. [65] developed a method in which trachea from the donor was first placed in the forearm and then transplanted into the recipient. The advantages of this method were revascularization and epithelization to the posterior trachea, which was removed from the trachea and replaced with mucosa from the recipient. Immunosuppression was used for all recipients. In that clinical study, complications such as necrosis and tracheobronchomalacia were encountered, although the outcomes were relatively better than in previous studies. The limitations of this method included necrosis of the posterior membranous wall due to a lack of circumferential revascularization and a need for long-term immunosuppressive therapy, which should be making this approach inapplicable for recipients with malignant tumors. Despite these drawbacks, the method developed by Delaere is a more promising approach compared to other methods.

5.7   Tracheal Allograft

Cadaveric trachea, in which all cells and histocompatibility markers have been deleted by chemical treatment, has been transplanted to recipients with congenital, post-traumatic, and prolonged intubation stenosis [66]. As a result, 83% of recipients survived, and 66% showed no further airway problems after transplant. In the transplanted trachea, epithelial colonization was found in the lumen. Furthermore, this procedure was added to six more patients. Five patients survived, of whom three showed tracheomalacia, which required tracheal stent [67]. Another study using the same method showed a high rate of re-intervention, but this included minor procedures (granulation removal, tracheal dilation, endoscopic laser treatment, mitomycin application, stent placement, supraglottoplasty, and endoscopic arytenoidectomy, tonsillectomy, and adenoidectomy) as well as major procedures (laryngotracheoplasty and slide tracheoplasty) [68].

The limitations include a need for long-term tracheal stenting with the attending possible complications, development of serious tracheomalacia, and reduced colonization by ciliary epithelium. Tracheal allografts are unlikely to show perfect function as a template for regeneration of the complex tracheal structure because of the high morbidity rate.

5.8 Tissue Engineering

Tissue-engineered trachea in general includes development of a trachea in association with a three-dimensional matrix using stem cells. Use of a bioartificial nano-composite was described in 2011, reportedly incurring no major complications, and the patient remained asymptomatic and tumor-free as of 5 months after the surgery [69]. However, investigations by independent parties resulted in concerns regarding potential scientific misconduct associated with that case report. The authors described a need for stents to stabilize the airway [70], and epithelization of the graft remained unconfirmed in terms of the growth of epithelial structures [71]. Furthermore, the patient developed serious complications and died [70]. As a result, the applicability of the described methods to patients is thus highly questionable.

On the other hand, decellularized trachea appears to represent the best scaffold identified to date, because of the non-immunological extracellular matrix (ECM) lacking major histocompatibility complex class I (MHC-I) and class II (MHC-II) proteins and the decellularized trachea lacks these pro-angiogenic properties or has them [32, 72]. At the same time, the remaining matrix supports stem cell differentiation and regeneration without a requirement for biological additives [73, 74]. Even though determining how much cartilage should be decellularized before recellularization remains difficult, some studies have shown clinical applicability [65, 75].

Further, an anatomy-mimicking artificial trachea, designated "BIO-AIR-TUBE" and consisting of engineered cartilage rings and autologous fibrous connective tissue, has been reported [76]. This suggests the possibility of easily generating anatomy-mimicking cylinder-type airways.

6 Bio-3D-Printed Trachea

Most artificial airway organs still use scaffolds to maintain airway strength and stiffness, as described above. However, scaffolds for artificial organs show limitations such as risk of infection, irritation, reduced biocompatibility, and degradation over time [77, 78]. Immunosuppression for allograft transplantation also risks infection and malignant disease [79]. To address these issues, researchers have selected and investigated various kinds of scaffold. To simplify the issues involved, artificial organs consisting of cells from the patient without scaffolds would be ideal.

So-called 3D bioprinting involves some promising technologies for artificial trachea, including digital light processing (DLP) 3D printers for bioprinting chondrocytes with silk-glycidyl methacrylate (GMA) [80], 3D bioprinting multi-layered scaffolds using polycaprolactone (PCL) and hydrogel with nasal epithelial and auricular cartilage cells [81], and another bioprinting technology to create scaffold-free tubular tissue from spheroids using the Regenova bio-3D printer-based system (Cyfuse Biomedical K.K., Tokyo, Japan) [82, 83].

DLP 3D printing systems use silk fibroin (SF) as a natural polymer fabricated with GMA (silk-GMA) and including chondrocytes. The method is as follows: as a pre-hydrogel, fabricated SF was chemically modified with GMA. Degummed silk was dissolved in 9.3 M lithium bromide, and the resulting solution reacted with GMA for 6 h at 60 °C. To remove salts, dialyzation was performed with distilled water at room temperature for 7 days, and then freeze-drying was performed. The DLP 3D bioprinting procedure using human, or rabbit-derived chondrocytes were mixed with silk-GMA, including lithium phenyl- (2,4,6-trimethylbenzoyl) phosphinate (LAP) as a photopolymer reagent. Chondrocyte-laden silk-GMA was printed in a layering type with a DLP printer. The liquid silk-GMA bio-ink with LAP hardens under ultraviolet exposure. Printed chondrocyte-laden silk GMA hydrogels were cultured for 1 week under in vitro conditions. Four weeks of in vitro cultivation of silk-GMA hydrogel ensures viability, proliferation, and differentiation to chondrogenesis for encapsulated cells. In vivo transplantation of artificial trachea to a partial defect in a rabbit model demonstrated new cartilage-like tissue and epithelium surrounding the transplanted silk-GMA hydrogel. This represents a promising method to create an artificial trachea with sufficient durability for transplantation. However, application has only been reported for partial defect of the trachea, not full-circumference replacement [80].

The process for another 3D bioprinting system using PCL and hydrogel is as follows. The 3D bioprinter (KIMM & Protek Korea, Daejeon, Korea) consists of four parts: a "screw pump system" for hydrogel printing, a "screw pump and heating controller" for PCL printing, a "tube fabrication controller" for tube shape formation, and a "motion controller." To fabricate the scaffold, PCL (catalog no. 440744; Sigma-Aldrich, MO, USA) polymer pellets were ejected through a heated nozzle to create the first layer. As the next step, a hydrogel comprising target cells and sodium alginate (Sigma-Aldrich) was used to print the second layer. The artificial trachea was constructed in a total of five layers: first layer, PCL (grid pattern, 5-mm diameter); second layer, hydrogel (cylindrical, 6-mm diameter) with epithelial cells; third layer, PCL (cylindrical, 6.5-mm diameter); fourth layer, hydrogel with chondrocytes (cylindrical, 7-mm diameter); and fifth layer, PCL (grid pattern, 8-mm diameter) [81]. This artificial trachea was successfully transplanted into partial resected trachea in situ for New Zealand rabbits. Again in that study, artificial trachea was transplanted into a partial defect of the trachea, not as a full-circumference replacement. For implanted cells, they used only two types: chondrocyte and epithelial cells. Furthermore, foreign materials were used, in the form of biodegradable polyester. These represent the key disadvantages of the method.

Artificial organs consisting of cells from the patient without scaffolds would be ideal from the perspectives of not needing immunosuppression due to a lack of immunoreaction, absence of infection due to the lack of foreign material, and no added risk of malignancy. In particular, infection is of central importance for the airway, which remains open to the air of the external environment. In this regard, foreign materials represent a problematic issue. We are trying to create a scaffold-free artificial trachea using the computer-controlled Regenova robotics system. This

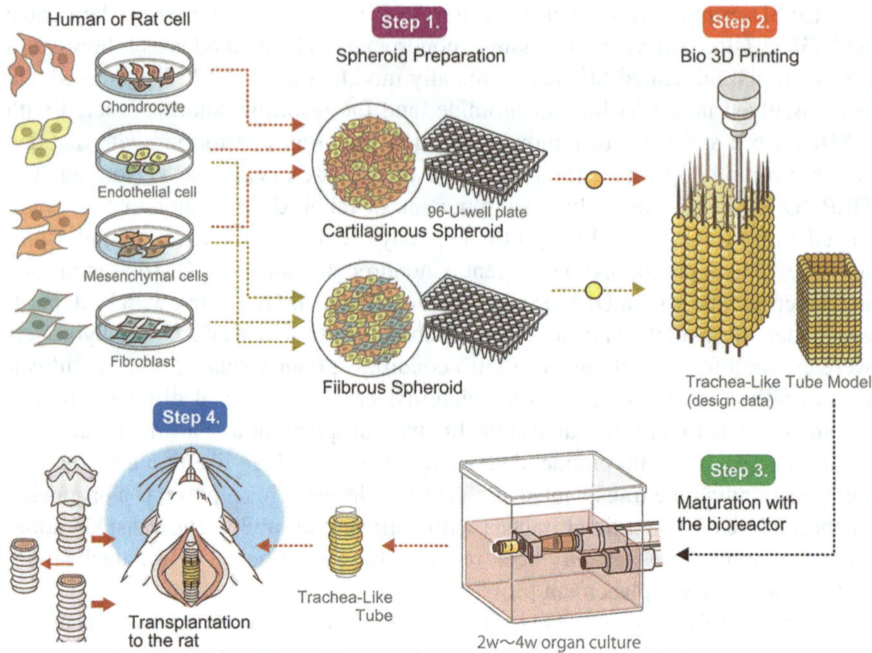

Fig. 2 The process of making artificial trachea using bio-3D printer system "Regenova." (Modified from Ref. [83]. Copyright Wiley-VCH Verlag GmbH & Co. KGaA. Reproduced with permission)

technology may permit the production of autologous 3D structures by isolating autologous cells and may allow optimal tracheal transplantation and regeneration without the use of foreign materials or immunosuppressants (Fig. 2).

In our first study, we attempted to generate neo-cartilaginous tubes and neo-fibrous tubes by co-culturing human chondrocytes, human mesenchymal stem cells (MSCs), human fibroblasts, and human umbilical vein endothelial cells (HUVECs). In addition, we attempted to build a structure more similar to native trachea, with suitable strength for rat transplant [83]. First, we addressed the following issues using human cells above: (1) what kind of cells should be used, (2) what kind of cell combination that would be useful, and (3) what kind of structure is useful and strong enough for transplantation. As a result, the trachea-like tube comprising cartilaginous and fibrous layers was stronger than more purely cartilaginous or fibrous tubes. Tensile strength of the trachea-like tube measured using Tissue Puller (DMT; Ann Arbor, MI, USA) was similar to that of a 9-week-old native rat trachea. In addition, the histologically trachea-like structure showed the formation of small capillaries comprising CD31-positive cells; vascular-like structures had thus been growing in the artificial trachea during in vitro culture. These were transplanted into rat trachea circumferentially in situ with stenting. The recipient rats survived more than 1 month.

Histologically, anti-CD31-positive capillary-like structures were found in the graft. Further, epithelization extended from native trachea even though full coverage of epithelization was not seen on the graft surface as of just 1 month after transplantation. This means that epithelization of transplanted trachea should not be expected until more than 1 month after transplantation. On the other hand, in that study, creation of cartilage was insufficient after transplantation. This may have been because the human chondrocytes used were not primary cultured chondrocytes, but instead had a partially fibroblastic character because of the rapid dedifferentiation during monolayer culture, resulting in a fibroblast-like phenotype producing less cartilage than primary cells would have [84]. In that study, the cartilaginous tube comprising chondrocytes alone offered insufficient strength for transplantation. Other cell types were thus needed to strengthen the structures. One promising cell source was MSCs. However, these cells produce cartilage after chondrogenic induction, which can lead to further hypertrophy and calcification. Furthermore, MSCs currently remain quite expensive [85]. Some studies have described co-culture of articular chondrocytes with MSCs, leading the MSCs to give rise to chondrocytes via interactions that maintain the chondrogenic phenotypes.

In a second study, to form a scaffold-free artificial trachea, we attempted to use primary cultured cells isolated from F344 rats, which include mesenchymal stem cells, chondrocytes, and endothelial cells [82]. Isolation of endothelial cells proved technically difficult, and we eventually used endothelial cells from rat lung microvessels (VEC Technologies, Rensselaer, NY, USA). In that study, structure strength was greater than that of trachea from 3-week-old rats after in vitro culture, despite being almost half that of trachea from 8-week-old rats. In vivo, cartilage was clearly found in the structure 23 days after transplantation, even though the chondrocytes formed clusters. Of course, epithelization was evident on the lumen surface, and anti-CD31-positive capillary-like structures were found in the graft wall. Furthermore, red blood cells were identified inside the capillary-like tube formations. As a result, rats transplanted with artificial trachea consisting of primary culture cells survived more than 1 year. However, the long-term (more than 3 months) survival rate may be about 40%. Other issues in this study included the instability of cartilage in the structure, reduced strength compared to native trachea in adult rats, and a lack of data on long-term outcomes without tracheal stenting.

As another issue, to make these scaffold-free structures at a human size using cells, large aggregations of cells would be necessary in the future. Endothelial cells are particularly important. In fact, previous studies have shown that HUVECs play an important role in such structures. However, HUVECs cannot be isolated from adult humans. As alternatives, cells differentiated from stem cells, especially induced pluripotent stem cells (iPSCs), are strong candidates for artificial organs. We compared the potentials of HUVEC and iPSC-derived endothelial cells in scaffold-free trachea [86]. No significant difference in tensile strength was identified. Histologically, small capillary-like tube formations comprising CD31-positive cells were observed in all groups. The numbers and diameters were significantly lower in the iPSC-derived endothelial cell group than in other groups. Glycosaminoglycan content was significantly lower in the iPSC-derived endothelial cell group than in the HUVEC group. Based on these data, some limitations appear associated with iPSC-derived endothelial cells compared to HUVECs in scaffold-free structures.

7 Future Perspectives

Scaffold-free artificial trachea seems likely to have a great future from the perspective of using recipient-derived cells, avoiding the need for immunosuppression, lacking foreign material, and meeting expectations for organ growth with age. However, we have only succeeded in transplanting scaffold-free artificial trachea into small animals. Therefore, numerous issues remain: (1) instability of cartilage created in the structure, (2) quicker creation of a larger volume of blood flow in the graft, (3) the need for numerous cells to create a human-sized trachea, (4) a lack of data on long-term outcomes without tracheal stenting, and (5) the need for an alternative cell source for HUVECs. Overall, this research is still only just over the starting line. Much more research is needed before we reach the goal of ideal organs (Fig. 3).

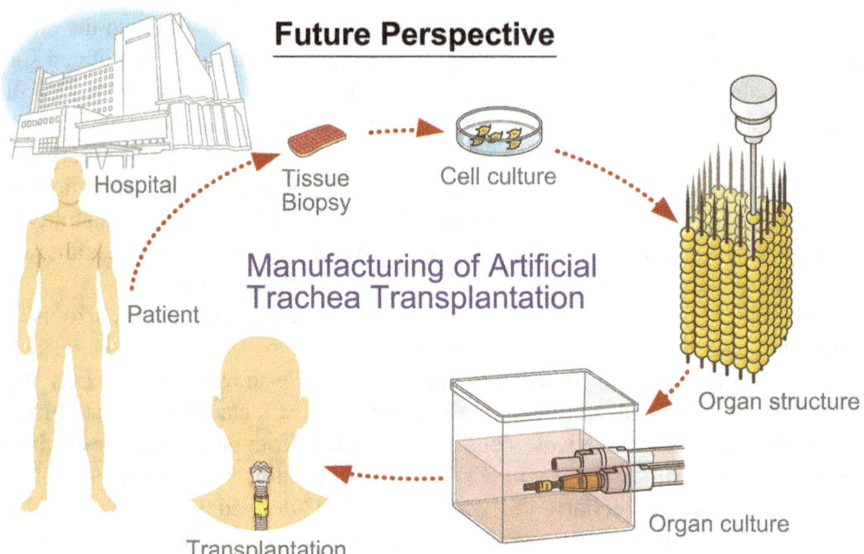

Fig. 3 Manufacturing process of artificial trachea for humans in the future

References

1. Mintz ML (2006) Disorders of the respiratory tract. Common challenges in primary care. Springer, Heidelberg, pp p11–p15
2. Furlow PW, Mathisen DJ (2018) Surgical anatomy of the trachea. Ann Cardiothorac Surg 7:255–260
3. Weibel ER (1963) Morphometry of the human lung. Academic, New York
4. Mercer RR, Russell ML et al (1994) Cell number and distribution in human and rat airways. Am J Respir Cell Mol Biol 10:613–624
5. Weibel ER (1997) Design and morphometry of the pulmonary gas exchanger. In: Crystal RG, West JB, Weibel ER et al (eds) The lung: scientific foundations, 2nd edn. Lippincott-Raven, Philadelphia, pp 1147–1157
6. Manninen MP, Antila PJ et al (1991) Occurrence of tracheal carcinoma in Finland. Acta Otolaryngol 111:1162–1169
7. Licht PB, Friis S et al (2001) Tracheal cancer in Denmark: a nationwide study. Eur J Cardiothorac Surg 19:339–345
8. Gelder CM, Hetzel MR (1993) Primary tracheal tumours: a national survey. Thorax 48:688–692
9. Bhattacharyya N (2004) Contemporary staging and prognosis for primary tracheal malignancies: a population-based analysis. Otolaryngol Head Neck Surg 131:639–642
10. Gilbert JB, Mazzarella LA et al (1949) Primary tracheal tumours in infants and children. J Pediatr 35:63–69
11. Desai DP, Holinger LD et al (1998) Tracheal neoplasm in children. Ann Otol Rhinol Laryngol 107:790–796
12. Perelman MI, Koroleva NS (1987) Primary tumors of the trachea. In: Grillo HC, Eschapasse H (eds) Major challenges, international trends in general thoracic surgery, vol 2. Saunders, Philadelphia, pp 91–106
13. Beheshti J, Mark EJ et al (2004) Epithelial tumor of the trachea. In: Grillo HC (ed) Surgery of the trachea and bronchi. BC Decker, London, pp 73–85
14. Beheshti J, Mark EJ (2004) Mesenchymal tumor of the trachea. In: Grillo HC (ed) Surgery of the trachea and bronchi. BC Decker, London, pp 86–97
15. Grillo HC (2004) Primary tracheal tumours. In: Grillo HC (ed) Surgery of the trachea and bronchi. BC Decker, London, pp p208–p247
16. Ernst A, Carden K et al (2015) Tracheomalacia and tracheobronchomalacia in adults. In: Basow DS UpToDate Waltham
17. Carden KA, Boiselle PM et al (2005) Tracheomalacia and tracheobronchomalacia in children and adults: an in-depth review. Chest 127:984–1005
18. Ridge CA, O'Donnell CR et al (2011) Tracheobronchomalacia: current concepts and controversies. J Thorac Imaging 26:278–289
19. Buitrago DH, Wilson JL et al (2017) Current concepts in severe adult tracheobronchomalacia: evaluation and treatment. J Thorac Dis 9:E57–E66
20. Mathisen DJ (1996) Tracheal tumours. Chest Surg Clin N Am 6:875–898
21. Mathisen DJ, Grillo HC (1989) Endoscopic relief of malignant airway obstruction. Ann Thorac Surg 48:469–475
22. Grillo HC, Mathisen DJ (1990) Primary tracheal tumours: treatment and results. Ann Thorac Surg 49:69–77
23. Pearson FG, Todd TRJ et al (1984) Experience with primary neoplasms of the trachea and carina. J Thorac Cardiovasc Surg 88:511–518
24. Pearson F, Brito-Filomeno L et al (1986) Experience with partial cricoid resection and thyrotracheal anastomosis. Ann Otol Rhinol Laryngol 95:582–585
25. Grillo HC (2003) Development of tracheal surgery: a historical review. Part 1: techniques of tracheal surgery. Ann Thorac Surg 75:610–619
26. Dedo H, Fishman N (1969) Suprahyoid release for tracheal stenosis. Ann Otol Rhinol Laryngol 78:285–291

27. Montgomery WW (1974) Suprahyoid release for tracheal stenosis. Arch Otolaryngol 99:255–259
28. Grillo HC (2003) Development of tracheal surgery: a historical review. Part 2: treatment of tracheal diseases. Ann Thorac Surg 75:1039–1047
29. Belsey R (1950) Resection and reconstruction of the intrathoracic trachea. Br J Surg 38:200–205
30. Rose KG, Sesterhenn K et al (1979) Tracheal allotransplantation in man. Lancet 1:433
31. Etienne H, Fabre D et al (2018) Tracheal replacement. Eur Respir J 51:1–9
32. Elliott MJ, De Coppi P et al (2012) Stem-cell-based, tissue engineered tracheal replacement in a child: a 2-year follow-up study. Lancet 380:994–1000
33. Jackson TL, O'Brien EJ et al (1950) The experimental use of homogenous tracheal transplants in the restoration of continuity of the tracheobronchial tree. J Thorac Surg 20:598–612
34. Scherer MA, Ascherl R et al (1986) Experimental biosynthetic reconstruction of the trachea. Arch Otorhinolarynol 243:215–223
35. Grillo HC (2002) Tracheal replacement: a critical review. Ann Thorac Surg 73:1995–2004
36. Spaggiari L, Calabrese LS et al (2005) Successful subtotal tracheal replacement (using a skin/omental graft) for dehiscence after a resection for thyroid cancer. J Thorac Cardiovasc Surg 129:1455–1456
37. Delaere PR, Vranckx JJ et al (2012) Learning curve in tracheal allotransplantation. Am J Transplant 12:2538–2545
38. Shinohara H, Yuzuriha S et al (2004) Tracheal reconstruction with a prefabricated deltopectoral flap combined with costal cartilage graft and palatal mucosal graft. Ann Plast Surg 53:278–281
39. Maciejewski A, Szymczyk C et al (2009) Tracheal reconstruction with the use of radial fore-arm free flap combined with biodegradative mesh suspension. Ann Thorac Surg 87:608–610
40. Yu P, Clayman GL et al (2006) Human tracheal reconstruction with a composite radial forearm free flap and prosthesis. Ann Thorac Surg 81:714–716
41. Fabre D, Fadel E et al (2015) Autologous tracheal replacement for cancer. Chin Clin Oncol China 4:46
42. Fabre D, Singhal S et al (2009) Composite cervical skin and cartilage flap provides a novel large airway substitute after long-segment tracheal resection. J Thorac Cardiovasc Surg 138:32–39
43. Fabre D, Kolb F et al (2013) Successful tracheal replacement in humans using autologous tis-sues: an 8-year experience. Ann Thorac Surg 96:1146–1155
44. Doss AE, Dunn SS et al (2007) Tracheal replacements: part 2. ASAIO J 53:631–639
45. Toomes H, Mickisch G et al (1985) Experiences with prosthetic reconstruction of the trachea and bifurcation. Thorax 40:32–37
46. Deslauriers J, Ginsberg RJ et al (1975) Innominate artery rupture. A major complication of tracheal surgery. Ann Thorac Surg 20:671–677
47. Gaissert HA, Grillo HC et al (2003) Complication of benign tracheobronchial strictures by self-expanding metal stents. J Thorac Cardiovasc Surg 126:744–747
48. Cull DL, Lally KP et al (1990) Tracheal reconstruction with polytetrafluoroethylene graft in dogs. Ann Thorac Surg 50:899–901
49. Grillo HC (1970) Surgery of the trachea. Curr Probl Surg:3–59
50. Guijarro Jorge R, Sanchez-Palencia Ramos A et al (1990) Experimental study of a new porous tracheal prosthesis. Ann Thorac Surg 50:281–287
51. Jacobs JR (1988) Investigations into tracheal prosthetic reconstruction. Laryngoscope 98:1239–1245
52. Neville WE, Bolanowski PJ et al (1990) Clinical experience with the silicone tracheal proth-esis. J Thorac Cardiovasc Surg 99:604–613
53. Pearson FG, Henderson RD et al (1968) The reconstruction of circumferential tracheal defect with a porous prosthesis. J Thorac Cardiovasc Surg 55:605–616
54. Neville WE, Bolanowski PJ et al (1976) Prosthetic reconstruction of the trachea and carina. J Thorac Cardiovasc Surg 72:525–538

55. Behrend M, Kluge E et al (2006) On the use of unsealed polypropylene mesh as tracheal replacement. ASAIO J 52:328–333
56. Maziak DE, Todd TR et al (1996) Adenoid cystic carcinoma of the airway: thirty-two-year experience. J Thorac Cardiovasc Surg 112:1522–1531
57. Debry C, Dupret-Bories A et al (2014) Laryngeal replacement with an artificial larynx after total laryngectomy: the possibility of restoring larynx functionality in the future. Head Neck 36:1669–1673
58. Debry C, Vrana NE et al (2017) Implantation of an artificial larynx after total laryngectomy. N Engl J Med 376:97–98
59. Liu Y, Lu T et al (2016) Collagen-conjugated tracheal prosthesis tested in dogs without omental wrapping and silicone stenting. Interact Cardiovasc Thorac Surg 23:710–715
60. Seguin A, Radu D et al (2009) Tracheal replacement with cryopreserved, decellularized, or glutaraldehyde-treated aortic allografts. Ann Thorac Surg 87:861–867
61. Tsukada H, Ernst A et al (2010) Tracheal replacement with a silicone-stented, fresh aortic allograft in sheep. Ann Thorac Surg 89:253–258
62. Wurtz A, Porte H et al (2010) Surgical technique and results of tracheal and carinal replacement with aortic allografts for salivary gland-type carcinoma. J Thorac Cardiovasc Surg 140:387–393
63. Levashov Yu N, Yablonsky PK et al (1993) One-stage allotransplantation of thoracic segment of the trachea in a patient with idiopathic fibrosing mediastinitis and marked tracheal stenosis. Eur J Cardiothorac Surg 7:383–386
64. Klepetko W, Marta GM et al (2004) Heterotopic tracheal transplantation with omentum wrapping in the abdominal position preserves functional and structural integrity of a human tracheal allograft. J Thorac Cardiovasc Surg 127:862–867
65. Delaere P, Vranckx J et al (2010) Tracheal allotransplantation after withdrawal of immunosuppressive therapy. N Engl J Med 362:138–145
66. Jacobs JP, Elliott MJ et al (1996) Pediatric tracheal homograft reconstruction: a novel approach to complex tracheal stenoses in children. J Thorac Cardiovasc Surg 112:1549–1558
67. Jacobs JP, Quintessenza JA et al (1999) Tracheal allograft reconstruction: the total North American and worldwide pediatric experiences. Ann Thorac Surg 68:1043–1051
68. Propst EJ, Prager JD et al (2011) Pediatric tracheal reconstruction using cadaveric homograft. Arch Otolaryngol Head Neck Surg 137:583–590
69. Jungebluth P, Alici E et al (2011) Tracheobronchial transplantation with a stem-cell-seeded bioartificial nanocomposite: a proof-of-concept study. Lancet 378:1997–2004
70. Claesson-Welsh L, Hansson GK, Royal Swedish Academy of Sciences (2016) Tracheobronchial transplantation: the Royal Swedish Academy of Sciences' concerns. Lancet 387:942
71. Cyranoski D (2014) Investigations launched into artificial tracheas. Nature 516:16–17
72. Jungebluth P, Moll G et al (2012) Tissue-engineered airway: a regenerative solution. Clin Pharmacol Ther 91:81–93
73. Benders KE, van Weeren PR et al (2013) Extracellular matrix scaffolds for cartilage and bone regeneration. Trends Biotechnol 31:169–176
74. Schwarz S, Elsaesser AF et al (2015) Processed xenogenic cartilage as innovative biomatrix for cartilage tissue engineering: effects on chondrocyte differentiation and function. J Tissue Eng Regen Med 9:E239–E251
75. Delaere P, Van Raemdonck D (2016) Tracheal replacement. J Thorac Dis 8:S186–S196
76. Komura M, Komura H et al (2019) Fabrication of an anatomy-mimicking BIO-AIR-TUBE with engineered cartilage. Regen Ther 11:176–181
77. Jungebluth P, Haag JC et al (2014) Tracheal tissue engineering in rats. Nat Protoc 9:2164–2179
78. Kelm JM, Lorber V et al (2010) A novel concept for scaffold-free vessel tissue engineering: self-assembly of microtissue building blocks. J Biotechnol 148:46–55
79. Hysi I, Kipnis E et al (2015) Successful orthotopic transplantation of short tracheal segments without immunosuppressive therapy. Eur J Cardiothorac Surg 47:e54–e61
80. Hong H, Seo YB et al (2019) Digital light processing 3D printed silk fibroin hydrogel for cartilage tissue engineering. Biomaterials 232:119679

81. Park JH, Yoon JK et al (2019) Digital light processing 3D printed silk fibroin hydrogel for cartilage tissue engineering. Experimental tracheal replacement using 3-dimensional bioprinted artificial trachea with autologous epithelial cells and chondrocytes. Sci Rep 9:2103
82. Taniguchi D, Matsumoto K et al (2018) Scaffold-free trachea regeneration by tissue engineering with bio-3D printing. Interact Cardiovasc Thorac Surg 26:745–752
83. Machino R, Matsumoto K et al (2019) Replacement of rat tracheas by layered, trachea-like, scaffold-free structures of human cells using a bio-3D printing system. Adv Healthc Mater 8:e1800983
84. Dell'Accio F, De Bari C et al (2001) Molecular markers predictive of the capacity of expanded human articular chondrocytes to form stable cartilage in vivo. Arthritis Rheum 44:1608–1619
85. Dickhut A, Pelttari K et al (2009) Calcification or dedifferentiation: requirement to lock mesenchymal stem cells in a desired differentiation stage. J Cell Physiol 219:219–226
86. Taniguchi D, Matsumoto K et al (2020) Human lung microvascular endothelial cells as potential alternatives to human umbilical vein endothelial cells in bio-3D-printed trachea-like structures. Tissue Cell 63:101321

Cardiac Tissue Creation with the Kenzan Method

Hiroshi Matsushita, Vivian Nguyen, Katherine Nurminsky,
and Narutoshi Hibino

Abstract There is currently an increasing demand for tissue-engineered cardiac patch creation. These techniques can be broadly classified into two categories based on the use, or lack thereof, of supporting biomaterials: also known as "scaffolds." The scaffold-free method uses spheroids as building blocks and depends on the spheroids to produce their own extracellular matrix, which in turn directs the spheroids' growth and development. This method creates patches with high biocompatibility, low excitability of immune responses, and integrated mechanical contractibility and electrical conductivity, in contrast with the scaffold dependent method. The Kenzan method is the most powerful and versatile technique available to create cardiac patches without scaffolds. In this paper, we reviewed cardiac patch creations with the Kenzan method and compared the patches with scaffold-based ones.

Keywords Cardiac patch · Scaffold-free · Spheroids

1 Introduction

Heart disease, also known as cardiovascular disease (CVD), is a type of disease that affects the heart and/or blood vessels of a person. It is the leading cause of death in the world and is expected to account for >23.6 million deaths by 2030 [1]. Current medical therapy approaches to heart disease only mitigate symptoms and modestly prolong life, but they do not fix the fundamental loss of functional heart tissue.

Heart transplantation can replace damaged hearts; however, the number of donors is limited, and not all patients are eligible for the procedure. For these

H. Matsushita · V. Nguyen · K. Nurminsky · N. Hibino (✉)
The University of Chicago Biological Sciences, Chicago, IL, USA
e-mail: nhibino@surgery.bsd.uchicago.edu

© Springer Nature Switzerland AG 2021

K. Nakayama (ed.), *Kenzan Method for Scaffold-Free Biofabrication*,
https://doi.org/10.1007/978-3-030-58688-1_8

reasons, a large effort has been made in researching cardiovascular tissue engineering methods including three-dimensional (3D) printing and bioprinting [2, 3].

Cardiovascular tissue engineering techniques can be broadly classified into two categories based on the use, or lack thereof, of supporting biomaterials – also known as "scaffolds." Biomaterial-dependent techniques seek to create a scaffold that will properly deliver cells and direct their growth, but these techniques suffer from several limitations such as rejection from the body (immunogenic response), not mimicking the native myocardium, and rapid degradation [4]. Several limitations of scaffold-dependent tissue engineering can be avoided by using biomaterial/scaffold-free methods [5]. Scaffold-free methods use spheroids as a building block, and they depend on the spheroids to produce their own extracellular matrix and to direct their own growth and development [4]. Currently, biomaterial-dependent and biomaterial-free techniques struggle to create vascularized tissue patches, with most engineered constructs being around 200 μm in thickness due to the limit of free oxygen diffusion in living tissues. It is difficult to provide innervation to constructs, and the lack of microvascular perfusion is a major roadblock to creating larger tissue patches [6].

This overview seeks to contextualize the field of 3D cardiovascular tissue creation and to focus on the Kenzan method, a versatile and powerful biomaterial-free fabrication technique that combines the advantages from biomaterial-dependent fabrication techniques with the advantages of biomaterial-free fabrication techniques.

2 Biomaterial-Dependent Creation of 3D Cardiovascular Tissue

Several biomaterials have been applied to cardiac tissue creation. In general, the goal of biomaterial-dependent techniques is to create a scaffold that will support and direct cell growth [7]. Besides preferred biocompatibility and low excitability of immune responses, these biomaterials require multiple criteria to be carefully considered for optimal tissue function, including physical properties of the polymer (e.g., strength and elasticity), degradation rates, and host response [8]. The appropriate biomaterial for cardiac tissue engineering should be as unique as the heart itself; it should be highly flexible, elastic, and capable of enduring millions of contraction cycles while supporting the seeded cell viability and differentiated phenotype both in vitro and in vivo [7]. This section will review applications of different biomaterials to 3D cardiovascular tissue creation. These materials can be divided into three categories based on their origin: natural, synthetic, and hybrid.

Natural scaffolds are made from materials generally found in the extracellular matrix (ECM), such as collagen, fibrin, or hyaluronan. These materials are both biocompatible and biomimetic to the natural ECM, therefore facilitating cell growth and proliferation. Chachques et al. implanted a cell-seeded collagen scaffold grafted onto infarcted ventricles [9]. In 15 patients (aged 54.2 ± 3.8 years) presenting LV (Left ventricle) post-ischemic myocardial scars and with indications for a single coronary artery bypass graft OP-CABG, autologous mononuclear bone marrow cells (BMC) were implanted in the scars during surgery. NYHA classification symptoms, LV end-diastolic, LV filling deceleration time, and LV EF (Ejection fraction) were improved. Ye et al. created IGF (insulin growth factor)-1-containing fibrin scaffold patches using human-induced pluripotent stem cells (hiPSCs), hiPSC-derived ECs (hiPSC-ECs), and smooth muscle cells (hiPSC-SMCs). They then implanted the patches into porcine models of acute myocardial infarction (MI), which showed improvement of LV EF and decrease of infarct size and regional wall stress [10].

In contrast to natural scaffolds, synthetic scaffolds are easily produced on a large scale, have a relatively low cost of production, and have easily customizable architecture and functionality [11]. Synthetic polymers, such as polyglycerol sebacate (PGS), polyethylene glycol (PEG), polyglycolic acid (PGA), poly-L-lactide (PLA), poly(lactide-co-glycolide) (PLGA), polyvinyl alcohol (PVA), polycaprolactone, polyurethanes, and poly(N-isopropylacrylamide) are being considered for developing scaffolds of cardiac patches [12]. Rai et al. effectively created a cardiac patch with polyglycerol sebacate (PGS) [13]. Iyer et al. developed a microchannel created scaffold with photopolymerization of polyethylene glycol [14]. However, synthetic scaffolds lack the biological cues and biomimetic properties natural scaffolds have, leading to poor cell growth and recognition as well as poor biocompatibility. Synthetic scaffold degradation by-products may also be toxic to the surrounding tissue and alter the local environment, which affects cell behavior and survival, and leads to tissue damage and inflammatory reactions [11].

Both natural and synthetic scaffolds have their advantages and disadvantages, and hybrid scaffolds seek to combine the superior functionality and biocompatibility of natural scaffolds with the strength, durability, and tunability of synthetic polymers. Heydarkhan-Hagvall et al. showed the effectiveness of gelatin/PCL blended scaffold [15]. Sreerekha et al. showed scaffold created with fibrin and poly(lactide-co-glycolide) formed a suitable environment to enhance the differentiation of mesenchymal stem cells into cardiomyocytes [16].

Cardiac tissue requires integrated electrical conductivity and specific mechanical characteristics for the contraction. The tissue's high oxygen consumption requires internal vasculature development. Besides the criteria mentioned before, one of the key requirements of biomaterials is to produce vascular networks within the structure. Most of the vasculatures produced in the cardiac patch are limited to ~200 um due to the diffusion limit. It is essential to overcome this limitation, especially for large cardiac patch creation.

3 Biomaterial-Free Creation of 3D Cardiovascular Tissue

Biomaterial-free creation of 3D cardiovascular tissue seeks to use the self-assembling nature of cells in order to create tissue patches. Spheroids make up the basis of this technique; spheroids are capable of forming their own extracellular matrix and can be assembled to form certain morphologies [4]. Biomaterial-free techniques work by lacing together preformed spheroids, which produce their own ECM which then supports cell growth and development. Spheroids then fuse into a tissue construct, which is removed from the device and cultivated further until the tissue displays functionality related to cardiac performance.

Several different methods have been used for cardiovascular TE, including culturing spheroids together by hand, the net mold method, and the Kenzan method [4, 17]. This section will focus on the Kenzan method, which combines advantages from scaffold-based techniques with the advantages of spheroids.

3.1 Spheroids for Cardiovascular Tissue Engineering

During embryonic development, cells undergo biological self-assembly to form complex tissues with 3D architecture and cell-to-cell connections that are mandatory for maintaining intracellular functions. The ECM provides biomechanical and biochemical cues that determine cell fate, proliferation, and homeostasis. Spheroids are thought to mimic these physiological conditions, and they are believed to be the key to building 3D tissue [18]. Spheroids have enhanced cell viability, morphology, polarization, and increased activity and physiological metabolic function. They also have high angiogenic and vasculogenic potential [19]. A co-culture of cell aggregates with endothelial cells (HUVECs) produced vasculature structures within the cardiac spheroids, followed by angiogenic sprouting into the spheroids [20]. Spheroids have been used for many different applications, and they are a promising path for 3D cardiovascular tissue engineering.

One of the first published uses of spheroids for 3D cardiovascular tissue engineering focused on the self-assembly nature of spheroids [4]. In this study, Noguchi et al. focused on using spheroids in order to combat the disadvantages of scaffold-based tissue-engineering approaches. They made 1000 cell cardiac spheroids out of cardiomyocytes, endothelial cells, and fibroblasts. These spheroids were positioned by hand on a low-attachment 100 mm plate in the desired shape by gently rotating the plate several times. After 24 h, the spheroids self-organized to form cardiac patches under static conditions. Unfortunately, this approach did not succeed in fabricating thicker myocardial tissue. The patch was not synchronically contracting with the native heart in in vivo models, and the patches will need to be scaled up in the future. This approach also demonstrated the struggles of using the spheroid-based method, as it was difficult to manipulate spheroids, a large number of cells was needed, and the patch lacked robustness.

Other spheroid-based methods developed after Noguchi et al.'s study, such as the net mold method [17]. However, all of these methods struggled with the precise placement of spheroids and the consistent development of a tissue patch. Eventually, the Kenzan method was developed to try and combat these difficulties.

3.2 *The Kenzan Method*

The main difficulty with previous spheroid-based approaches has been issues in controlling the distance between spheroids [4]. Spheroids placed too far apart do not fuse, and spheroids placed too close together face high cell death [17, 21]. The Kenzan method combats this difficulty by placing spheroids in a pre-designed pattern, in which spatial organization and placement are precisely controlled. In addition to this, the grid configurations are flexible, which allows for cardiac patches of various geometries (e.g., tubes, rings, etc.) and thicknesses (number of layers). The tissues produced from the Kenzan method are more uniform at the macroscopic and microscopic level than past tissues made of random self-assembly of spheroids.

The current method of the Kenzan technique starts with the preassembly of cells into spheroids. These spheroids are transferred into non-adhesive round-bottomed 96-well tissue cultures plates. They are then picked up by a nozzle connected to a mobile arm and are robotically "impaled" onto microneedles as temporary support. The desired 3D geometry is prepared ahead of time by using the 3D design software, and the robotic arm places the spheroids on the exact spatial coordinates according to a pre-specified 3D design in a sterile environment. Once bioprinted, the cardiac patch matures for 72 h in culture with the needles in place while on a shaker at 150 rpm in the incubator. During this time, the spheroids fuse, and afterward the needles are removed and the cardiac patches are cultured further to allow the voids to seal and for maturation.

One of the benefits of the Kenzan method is that it produces a robust cardiac patch with a non-immunogenic and functional 3D extracellular matrix. Ong et al. reported in vitro and in vivo estimates of cardiac patches created with this method [21]. The group used human-induced pluripotent stem cell-derived cardiomyocytes (hiPSC-CMs), fibroblasts (FB), and endothelial cells (EC) for spheroid creation. For this method, a high quality of the spheroids was essential; the researchers therefore optimized the combination of these cells for spheroids and patch creation by comparing three groups with three different ratios of these cells (CM/FB/EC = 70:15:15, 70:0:30, 45:40:15). They also attempted to find the optimal number of cells. 33,000 cells per spheroids and CM/FB/EC = 70:15:15 group were chosen as optimal conditions. These were determined in light of the beating ratio, cell diameter, cell viability, and electrophysiological studies of the patch, found using optical mapping. Cardiac patches created by this group were implanted into the left ventricles of rat hearts and analyzed with histological studies. These patches showed in vivo engraftment and vascularization.

Yeung et al. evaluated the in vivo regenerative potential of these 3D bioprinted cardiac patches [22]. They created 3.6 mm × 3.6 mm cardiac patches with spheroids using 33,000 cells per a spheroid, co-cultured with three cell lines (hiPSC-CM/FBs/EC, 70:15:15). These patches were implanted into ischemic areas of rat hearts that were induced by ligation of the left anterior descending artery. They were compared to the control group which had no patch implantations. The potential for regeneration was measured 4 weeks after the surgery with histology and echocardiography. The average vessel counts within the infarcted area were higher, the scar area was significantly smaller, and cardiac function measured by echography showed a trend of improvement compared to the control group.

The Kenzan method is still limited in building thick tissues due to a lack of organized vasculature. The cell number, ratio of cell types, and arrangement of spheroids need to be considered. The cardiac patches are made from the cells secreting a limited amount of ECM, which causes weak mechanical properties.

3.3 Future Directions for the Kenzan Method

The Kenzan method has been used to print small diameter vascular grafts, tracheal and urethral tubes, neural bridges, and liver buds. It has also been used to print functional cardiac patches. The method has advantages and disadvantages, as shown in the table. Using this method, we can create cardiac tissues that have high biocompatibility, mechanical contractibility, and electrical conductivity. Furthermore, with this method, we do not need to be concerned about the excitability of immune responses that can be elicited by biomaterials. On the other hand, this method requires specific machines and the creation of high-quality spheroids. Mechanical strength of the patch was variable depending on cell types, and engraftment was variable. Vascularization inside of the patch is still limited when building thick structure. In the future, it would be beneficial to explore creating multi-layer spheroid patches and to try developing vasculature. We could potentially combine The Kenzan method with scaffold-dependent methods or other spheroid-based methods to develop vasculature and better mechanical strength.

	Kenzan	Biomaterial dependent creation
	High biocompatibility	Non dependence on spheroid
	Integrated mechanical contractibility	Low cost and effort
	Integrated electrical conductivity	High durability
Advantage	Low elicitability of immune response	High plasticity
	Requirement for a large number of cells	Low biocompatibility
	Requirement for specific printing system	variable contractibility
	Dependence on quality of spheroid	Variable electrical conductivity
Disadvantage	Low durability	High elicitability of immune response

4 Conclusion

CVD affects a large amount of population. Current treatments only mitigate symptoms, but we need to handle the loss of cells and heart/vasculature function. Researchers have begun to use tissue-engineered grafts to handle this cell loss. The Kenzan technique has proven to be powerful and versatile, and in the future, further developments or combinations with other techniques to create larger and stronger grafts could be promising.

References

1. Benjamin EJ et al (2019) Heart disease and stroke statistics; 2019 update: a report from the American Heart Association. Circulation 139(10):e56–e528
2. Giannopoulos AA et al (2016) Applications of 3D printing in cardiovascular diseases. Nat Rev Cardiol 13(12):701–718
3. Ong CS et al (2017b) Tissue engineered vascular grafts: current state of the field. Expert Rev Med Devices 14(5):383–392
4. Noguchi R et al (2016) Development of a three-dimensional pre-vascularized scaffold-free contractile cardiac patch for treating heart disease. J Heart Lung Transplant 35(1):137–145
5. Moldovan NI (2018) Progress in scaffold-free bioprinting for cardiovascular medicine. J Cell Mol Med 22(6):2964–2969
6. Elomaa L, Yang YP (2017) Additive manufacturing of vascular grafts and vascularized tissue constructs. Tissue Eng Part B Rev 23(5):436–450
7. Radisic M et al (2006) Biomimetic approach to cardiac tissue engineering: oxygen carriers and channeled scaffolds. Tissue Eng 12(8):2077–2091
8. Reis LA et al (2016) Biomaterials in myocardial tissue engineering. J Tissue Eng Regen Med 10(1):11–28
9. Chachques JC et al (2007) Myocardial assistance by grafting a new bioartificial upgraded myocardium (MAGNUM Clinical Trial): one year follow-up. Cell Transplant 16(9):927–934
10. Ye L et al (2014) Cardiac repair in a porcine model of acute myocardial infarction with human induced pluripotent stem cell-derived cardiovascular cells. Cell Stem Cell 15(6):750–761
11. Abbasian M et al (2019) Scaffolding polymeric biomaterials: are naturally occurring biological macromolecules more appropriate for tissue engineering? Int J Biol Macromol 134:673–694
12. Tallawi M et al (2015) Strategies for the chemical and biological functionalization of scaffolds for cardiac tissue engineering: a review. J R Soc Interface 12(108):20150254–20150254
13. Rai R et al (2013) Biomimetic poly(glycerol sebacate) (PGS) membranes for cardiac patch application. Mater Sci Eng C 33(7):3677–3687
14. Iyer RK et al (2009) Microfabricated poly(ethylene glycol) templates enable rapid screening of triculture conditions for cardiac tissue engineering. J Biomed Mater Res A 89A(3):616–631
15. Heydarkhan-Hagvall S et al (2008) Three-dimensional electrospun ECM-based hybrid scaffolds for cardiovascular tissue engineering. Biomaterials 29(19):2907–2914
16. Sreerekha PR et al (2013) Fabrication of electrospun poly (lactide-co-glycolide)-fibrin multiscale scaffold for myocardial regeneration in vitro. Tissue Eng Part A 19(7–8):849–859
17. Yang B et al (2019) A net mold-based method of biomaterial-free three-dimensional cardiac tissue creation. Tissue Eng Part C Methods 25(4):243–252
18. Laschke MW, Menger MD (2017) Life is 3D: boosting spheroid function for tissue engineering. Trends Biotechnol 35(2):133–144
19. Fennema E et al (2013) Spheroid culture as a tool for creating 3D complex tissues. Trends Biotechnol 31(2):108–115
20. Garzoni LR et al (2009) Dissecting coronary angiogenesis: 3D co-culture of cardiomyocytes with endothelial or mesenchymal cells. Exp Cell Res 315(19):3406–3418
21. Ong CS et al (2017a) Biomaterial-free three-dimensional bioprinting of cardiac tissue using human induced pluripotent stem cell derived cardiomyocytes. Sci Rep 7(1):4566
22. Yeung E et al (2019) Cardiac regeneration using human-induced pluripotent stem cell-derived biomaterial-free 3D-bioprinted cardiac patch in vivo. J Tissue Eng Regen Med 13(11):2031–2039

Scaffold-Free Autologous Cell-Based Vascular Graft for Clinical Application

Manabu Itoh

Abstract In clinical settings, conventional vascular prostheses have significant limitations with regard to thrombosis, infection, and biocompatibility, especially for small-caliber blood conduits. We have researched and developed a scaffold-free small-caliber vascular graft using a "Kenzan method Bio-3D Printer" to assemble multicellular spheroids to construct a three-dimensional structure predesigned on a computer system. This technique has enabled the production of tubular tissues only containing cells, without any exogenous scaffolds. Currently, we have been engaging in development that aims for the clinical application of autologous cell-based vascular graft made with this Bio-3D printer as a vascular access used for hemodialysis.

Keywords Vascular graft · Scaffold-free · Bio-3D printer · Tissue engineering · Cell-based

1 Introduction

In developed countries, the number of patients with serious diseases associated with poor circulation in major organs has been increasing, triggered by lifestyle changes, aging of population, and arteriosclerosis resulting from metabolic disorders, such as diabetes, hyperlipidemia, and hemodialysis [1, 2]. Even today, with spreading of endovascular treatment using catheters and stents, there is still a great need for

---

The author declares no conflicts of interest associate with this manuscript.

M. Itoh (✉)
Department of Thoracic and Cardiovascular Surgery, Faculty of Medicine, Saga University, Saga, Japan
e-mail: itomana@cc.saga-u.ac.jp

K. Nakayama (ed.), *Kenzan Method for Scaffold-Free Biofabrication*,
https://doi.org/10.1007/978-3-030-58688-1_9

surgery implanting a patient's own blood vessels or synthetic vascular prostheses, including revascularization for coronary artery disease and lower-limb arterial disease and vascular access reconstruction for hemodialysis [3, 4]. Existing small-caliber vascular prostheses made of foreign substances such as ePTFE (expanded polytetrafluoroethylene) have many problems to be solved in terms of anti-thrombogenicity, anti-infectivity, biocompatibility, etc. Therefore, a wide range of research has been implemented for the development of small-caliber vascular prosthesis using the approaches of regenerative medicine and tissue engineering [5, 6].

Yamanaka et al. reported the establishment of human iPS cells in 2007 [7]. In consequence, tissue regeneration using self-derived cells which have the capability of solving the immunorejection problem has become more real, with increasing expectations to regenerative medicine and drug discovery. As the means of implanting cells, various implantation approaches have been developed: an approach to insert cells directly to the target tissue or organ, an approach using aggregated cells (multicellular spheroids) rather than single cells, an approach using a cell sheet [8], and an approach to construct a three-dimensional structure. Cells generally form aggregates in tissues, in which proteins referred to as the extracellular matrix are present to fill the gaps of aggregates. Tissue engineering involves artificially creating such structures. In tissue engineering, tissue-derived biomaterials, such as collagen, and biodegradable polymers which are pharmaceutically approved as medical materials including polylactic acid were generally used as scaffolds of cells. However, scaffolds which are fundamentally foreign materials have the risk of inflammation and infection. Therefore, approaches have been taken to construct three-dimensional tissues composed of cells alone without using scaffolds in recent years [9].

We have researched and developed a scaffold-free vascular graft without artificial materials, using 3D bioprinting technology which has a completely different concept from that of conventional tissue engineering. After mentioning the problems of existing synthetic vascular prostheses, which are made of artificial materials, used in surgical revascularization, here we describe the current situation, challenges, and future perspective of the clinical development of scaffold-free cell-based vascular graft using 3D bioprinting technology.

2   Problems of Synthetic Vascular Prostheses Made of Artificial Materials

In 1952, Voorhees et al. reported the construction of an artificial vascular prosthesis made of Vinyon N (polyvinyl chloride fiber), which was the material used for military parachutes at the time [10]. This is the world's first artificial vascular prosthesis constructed from synthetic fiber. In 1954, DeBakey used an artificial graft made of polyester for replacement of an abdominal aortic aneurysm. Further, a polyester vascular prosthesis has showed stable long-term results in patency and durability to become mainstream of a large-caliber vascular prosthesis (inner diameter of 10 mm or larger). In recent years, it has also contributed significantly to the treatment of thoracic and abdominal aortic aneurysms as a stent graft which has a tube of metal net inside a polyester vascular prosthesis.

On the other hand, in a small-caliber vascular prosthesis (<6 mm), since the 1970s, an expanded polytetrafluoroethylene (ePTFE) vascular prosthesis to which proteins and cells are less prone to attach due to its high hydrophobicity has been introduced. To date, through refinement of layer structures and improvement of the coating technology, it has become widely used as a vascular prosthesis for bypass surgery for arteriosclerosis obliterans of the lower extremities, shunt surgery in a child with congenital heart disease (subclavian to pulmonary artery shunt), and hemodialysis [11]. However, a current small-caliber vascular prosthesis containing artificial materials regarded as foreign substances remains to have many problems in clinical practice, such as thrombosis, infection, narrowing and occlusion associated with intimal thickening (hyperplasia) of the anastomotic site, and not growing. The development of a small-caliber vascular prosthesis is still under development; therefore self-derived blood vessels from limited parts of the body, such as the internal thoracic artery, the gastroepiploic artery, the radial artery, and the great saphenous vein, are particularly extracted to be used as bypass grafts in coronary artery bypass grafting and distal bypass grafting for below-the-ankle of lower extremities. However, in some cases, surgical revascularization must be abandoned due to the absence of usable self-derived blood vessels, when self-derived blood vessels are already extracted at repeat surgery or due to the effects of high-grade stenosis at the origin of the subclavian artery, arteriosclerosis of the arterial graft, and varicose veins of the lower extremities.

In addition, the number of end-stage renal disease patients who require renal replacement therapy including maintenance hemodialysis is increasing to 2.6 million in the world in 2010. Furthermore, Liyanage et al. showed in their review that it would possibly more than double by 2030 [12]. In Japan, the number of end-stage renal disease patients is growing every year in conjunction with increasing diabetic nephropathy. As a result, the number of maintenance hemodialysis patients exceeded 330,000 cases in 2017, with the sale of approximately 12,000 hemodialysis vascular prostheses annually in Japan. However, there are problems that infection often spreads from the puncture site for blood transmission and removal and that thrombotic occlusion often occurs, with the primary patency rate of 35.4% to 64.5% 1 year after implantation, which is not a satisfactory result [13, 14]. Not limited to a vascular graft for hemodialysis, once a vascular prosthesis containing artificial materials is infected, it is prone to form biofilm, potentially worsening a condition to septic state unless the vascular prosthesis is completely removed. Moreover, not only in the acute phase following implantation but also in the chronic phase, the risk of infection via fungal invasion into the body due to dental caries, external injury, and repeat surgery, triggered by declining of the recipient's immunity, cannot be completely avoided.

3   Cell-Based Small-Caliber Vascular Graft Using Kenzan Method Bio-3D Printers

Due to the self-aggregation phenomenon, which is the basic function adherent cells originally have, cells fuse each other to form well-shaped spheres (spheroids) only when placed in a low attachment well plate. By combining different types of cells, it is possible to generate a hetero-spheroid. For example, a tubular tissue spheroid can be created from a mixture of three types of cells constituting native blood vessels (vascular endothelial cells, vascular smooth muscle cells, and fibroblasts). In addition, a cardiac spheroid and a liver tissue spheroid can also be created in accordance with the intended use [15–19]. Moreover, primarily in the field of stem cell such as iPS cells and ES cells, spheroids have been reported as in vitro-generated small organoids due to self-organization in three-dimensional culture in recent years [20]. In other words, a spheroid has an organizational structure modeling on tissue and organ culture environment in the body, with higher function of cells than the cells in general monolayer culture. A spheroid is currently utilized in various fields, such as new drug development as human tissue model, embryology, and cell biology.

A mold is used as the means of preparing a large three-dimensional structure using spheroids [21]; furthermore, Nakayama et al. have enabled a more complex three-dimensional structure to be finely constructed utilizing 3D-bioprinting technology which precisely constructs layers of spheroids, considering a spheroid as one unit. It is a robotics system which places multiple spheroids in arbitrary XYZ positions (Fig. 1), with the following actual way to construct a three-dimensional structure. First, three-dimensional design of a structure is prepared with the workstation computer. Second, the required number of arbitrary size of spheroids which are the smallest units of the

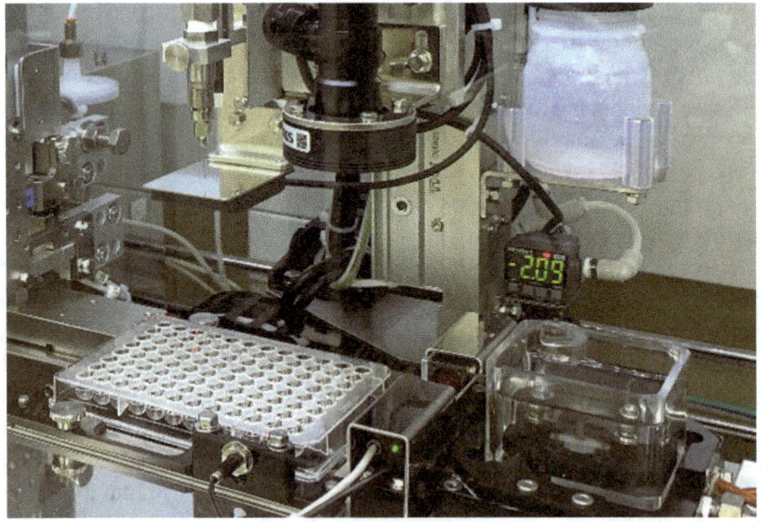

Fig. 1 Bio-3D printer (Regenova®, Cyfuse Biomedical K.K., Japan) (actually used)
A robotics system which is capable of constructing an arbitrary three-dimensional structure

structure to be created is prepared on a low attachment well plate. Spheroids are aspirated by a fine suction nozzle to be transferred robotically in accordance with the three-dimensional design and skewered into a needle array to be fixed. After a few days of culturing, layered spheroids are capable of retaining the structure after the removal of the needle array due to fusion and organization of the contiguous spheroids.

Cells cultured for expansion become to proliferate moderately and decrease requirements of oxygen and nutrients due to the formation of spheroids; thereby self-organization is promoted maintaining the survival rate due to the construction of three-dimensional layers after forming spheroids. In consequence, it is possible to create a highly condensed three-dimensional structure. In addition, in this Kenzan method using fine needles for temporal fixation, other molds which have been reported to be used in 3D-bioprinting, such as bio-paper and hydrogel, assuming what is called "printed paper" (scaffold material), are not required [22]. In other words, the Kenzan method has the advantage of the improvement of oxygen and nutrient circulation when a tissue is grown in size. Furthermore, due to being scaffold-free, it causes less inflammations and adverse effects associated with biomaterial degradation after transplantation and has the potential of self-organization which cells originally have (Fig. 2).

For further improvement of the dynamic strength as well as the physiological and molecular biological functions of cell-based tubular tissue, perfusion culture has been carried out for a certain period using a bioreactor. Successes in refinement of three-dimensional tissue maturation using the bioreactor and connection of structures inside the bioreactor made it possible to create a cell-based tubular tissue with an arbitrary diameter and length extending in the long axis direction (Fig. 3). We also succeeded in implanting to small animals (rats) [15], constructing a vascular graft from human dermal fibroblasts for hemodialysis, and implanting long-term to large animals (mini-pigs) (Fig. 4) [23].

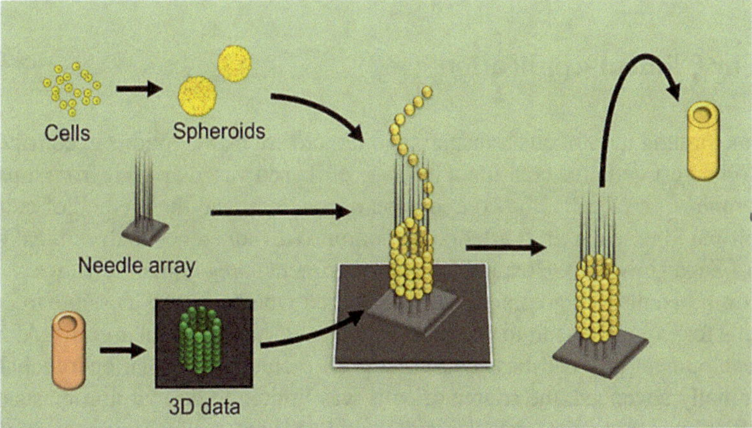

Fig. 2 The method to create a scaffold-free cell-based vascular graft (schema)
A robot makes layers of spheroids on the needle array. After culturing for a while, the needle array is removed with spheroids fusing each other to obtain a scaffold-free cell-based vascular graft

Fig. 3 A scaffold-free
cell-based vascular graft
(diameter of 5 mm, length
of 5 cm) (actually created)

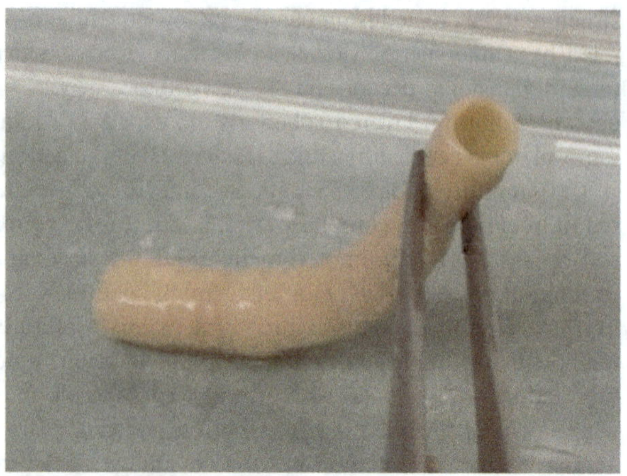

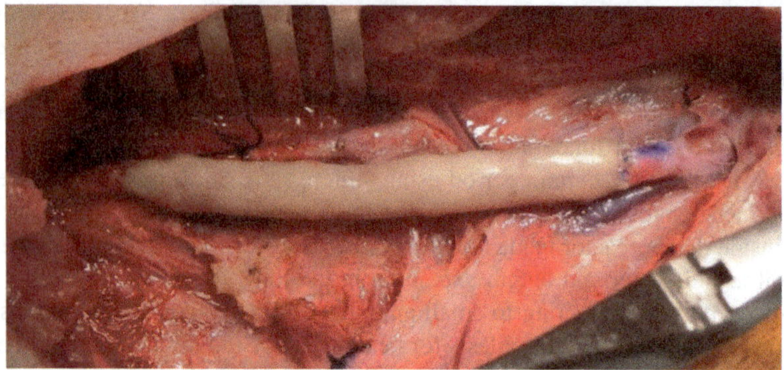

Fig. 4 Implantation to a large animal (actually conducted)
After neck arteriovenous shunt graft in a mini-pig

4 For Clinical Application

When assuming autologous transplantation, there are many options to consider: the
way to harvest cells for creating a human cell-based vascular graft from a patient,
the combination of cells, the composition of medium, and the method of culturing
the cell-based vascular graft after construction. With adherence to the ethical guidelines of Saga University Hospital for the experiment using human cells, we resected
cells from patients in agreement and considered whether it was possible to culture
the cells for expansion and to construct human cell-based vascular grafts. Although
the development of a cell-based vascular graft including vascular endothelial cells
was initially advanced, the source of cells was limited to dermal fibroblasts due to
the difficulty in collecting vascular endothelial cells and the invasion to patients. We
successfully developed human cell-based vascular grafts due to the isolation and the
culture for expansion of human dermal fibroblasts obtained from elderly patients

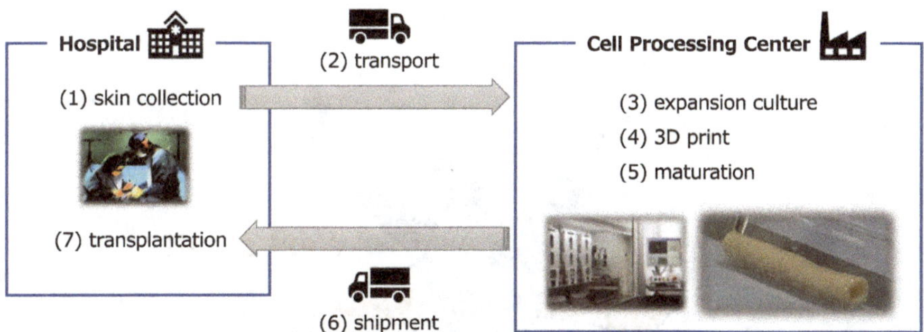

Fig. 5 Clinical study plan (schema)
(1), (2): Transportation of the resected skin from a patient in the hospital to a factory for manufacturing tissue-engineered medical products
(3), (4), (5): Isolation of dermal fibroblasts, culture for expansion, 3D print, and maturation of autologous cell-based vascular graft
(6), (7): Transportation of autologous cell-based vascular graft to the hospital and transplantation to the patient

with underlying diseases such as hypertension and hyperlipidemia and hemodialysis patients.

We are planning a clinical study to autograft a human cell-based vascular graft constructed by using a Bio-3D printer as vascular access reconstruction in hemodialysis (Fig. 5). This human cell-based vascular graft is created from patient cells alone without using artificial materials; therefore it may be expected to have utility in anti-infectivity and anti-thrombogenicity compared to synthetic vessels made of conventional artificial materials. In addition, the improvement of vascular access patency and the alleviation of patient discomfort owing to recurrent troubles are also expected. Moreover, measures to ensure safety are also the reason for targeting a vascular graft for hemodialysis, such as the relatively lower risk of life-threatening conditions due to troubles of a vascular graft for hemodialysis than that of a small-caliber vascular prosthesis required in other diseases and the regular observation which is enabled by three thrice-weekly visiting for dialysis.

5 Conclusion

In the only 70-year history of artificial blood vessels, the history of tissue-engineering accounts for less than half. If the construction of a functional small-caliber vascular graft made of cells alone is achieved due to our study, the clinical application of a vascular graft for hemodialysis which conventionally had problems of infection at the puncture site and thrombosis can be expected, followed by further applications to bypass graftings in the coronary artery and arteriosclerosis obliterans of the lower extremities in the future (Fig. 6). The functions of all organs in the human body are

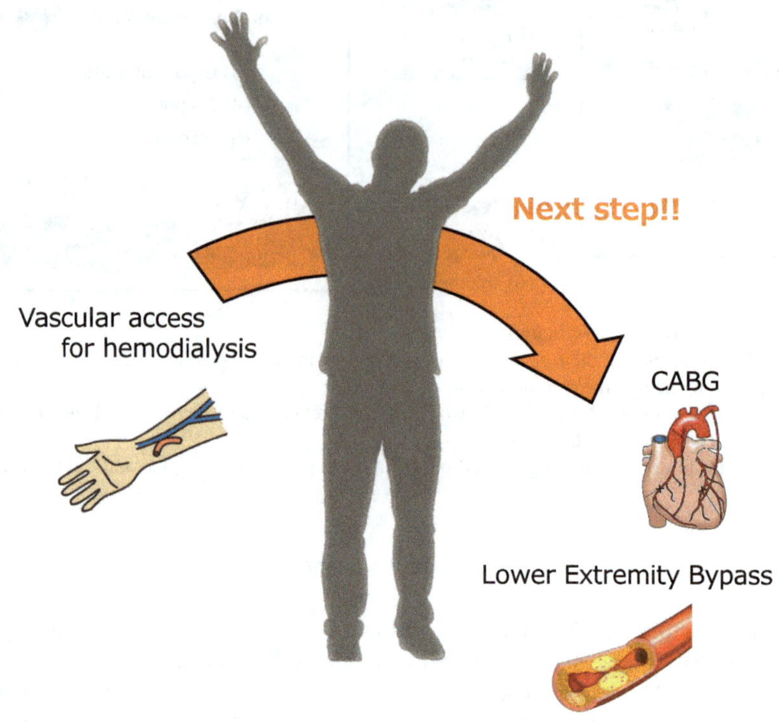

Fig. 6 Future perspective (schema)
The clinical application of a vascular graft for hemodialysis can be expected, followed by further applications, such as bypass grafting in arteriosclerosis obliterans of the lower extremities and coronary artery bypass grafting in the future

maintained by blood vessels; thus regenerative medicine and tissue-engineering approach may be expected to become a bridge not only to the development of artificial blood vessels but also to the regeneration of various organs.

References

1. Rafieian-Kopaei M, Setorki M, Doudi M et al (2014) Atherosclerosis: process, indicators, risk factors and new hopes. Int J Prev Med 5(8):927–946
2. Xu J, Murphy SL, Kochanek KD et al (2018) Deaths: final data for 2016. Natl Vital Stat Rep 67(5):1–76
3. Farkouh ME, Domanski M, Sleeper LA et al (2012) Strategies for multivessel revascularization in patients with diabetes. N Engl J Med 367(25):2375–2384
4. Doenst T, Haverich A, Serruys P et al (2019) PCI and CABG for treating stable coronary artery disease: JACC review topic of the week. J Am Coll Cardiol 73(8):964–976
5. Pashneh-Tala S, MacNeil S, Claeyssens F (2016) The tissue-engineered vascular graft-past, present, and future. Tissue Eng Part B Rev 22(1):68–100

6. Moldovan NI, Hibino N, Nakayama K (2017) Principles of the Kenzan method for robotic cell spheroid-based three-dimensional bioprinting. Tissue Eng Part B Rev 23(3):237–244
7. Takahashi K, Tanabe K, Ohnuki M et al (2007) Induction of pluripotent stem cells from adult human fibroblasts by defined factors. Cell 131(5):861–872
8. Kamao H, Mandai M, Okamoto S et al (2014) Characterization of human induced pluripotent stem cell-derived retinal pigment epithelium cell sheets aiming for clinical application. Stem Cell Rep 2(2):205–218
9. Nakayama K (2013) In vitro biofabrication of tissues and organs. In: Forgacs G, Sun W (eds) Biofabrication. William Andrew, New York, pp 1–21
10. Voorhees AB Jr, Jaretzki A 3rd, Blakemore AH (1952) The use of tubes constructed from vinyon "N" cloth in bridging arterial defects. Ann Surg 135(3):332–336
11. Samson RH, Morales R, Showalter DP et al (2106) Heparin-bonded expanded polytetrafluoroethylene femoropopliteal bypass grafts outperform expanded polytetrafluoroethylene grafts without heparin in a long-term comparison. J Vasc Surg 64(3):638–647
12. Liyanage T, Ninomiya T, Jha V et al (2015) Worldwide access to treatment for end-stage kidney disease: a systematic review. Lancet 385(9981):1975–1982
13. Guidelines of vascular access construction and repair for chronic hemodialysis (2011) J Jpn Soc Dial Ther 44:855–937
14. Huber TH, Carter JW, Carter RL et al (2003) Patency of autogenous and polytetrafluoroethylene upper extremity arteriovenous hemodialysis accesses: a systematic review. J Vasc Surg 38(5):1005–1011
15. Itoh M, Nakayama K, Noguchi R et al (2015) Scaffold-free tubular tissues created by a bio-3D printer undergo remodeling and endothelialization when implanted in rat aortae. PLoS One 10(9):e0136681
16. Mukae Y, Itoh M, Noguchi R et al (2018) The addition of human iPS cell-derived neural progenitors changes the contraction of human iPS cell-derived cardiac spheroids. Tissue Cell 53:61–67
17. Arai K, Murata D, Verissimo AR et al (2018) Fabrication of scaffold-free tubular cardiac constructs using a Bio-3D printer. PLoS One 13(12):e0209162
18. Kitsuka T, Itoh M, Amamoto S et al (2019) 2-Cl-C.OXT-A stimulates contraction through the suppression of phosphodiesterase activity in human induced pluripotent stem cell-derived cardiac organoids. PLoS One 14(7):e0213114
19. Yanagi Y, Nakayama K, Taguchi T et al (2017) In vivo and ex vivo methods of growing a liver bud through tissue connection. Sci Rep 7(1):14085
20. Takebe T, Enomura M, Yoshizawa E et al (2015) Vascularized and complex organ buds from diverse tissues via mesenchymal cell-driven condensation. Cell Stem Cell 16(5):556–565
21. Murata D, Tokunaga S, Tamura T et al (2015) A preliminary study of osteochondral regeneration using a scaffold-free three-dimensional construct of porcine adipose tissue-derived mesenchymal stem cells. J Orthop Surg Res 10:35
22. Norotte C, Marga FS, Niklason LE et al (2009) Scaffold-free vascular tissue engineering using bioprinting. Biomaterials 30:5910–5917
23. Itoh M, Mukae Y, Kitsuka T et al (2019) Development of an immunodeficient pig model allowing long-term accommodation of artificial human vascular tubes. Nat Commun 10(1):2244

Peripheral Nerve Regeneration Using Bio 3D Nerve Conduits

Ryosuke Ikeguchi, Tomoki Aoyama, Hirofumi Yurie, Hisataka Takeuchi, Sadaki Mitsuzawa, Maki Ando, Souichi Ohta, Takashi Noguchi, Shizuka Akieda, Koichi Nakayama, and Shuichi Matsuda

Abstract In peripheral nerve injuries with nerve defects, several surgical materials have been developed to bridge the nerve gaps. Autologous nerve grafts remain the gold standard method for their nerve regeneration capacities. In this chapter, we describe the basic mechanisms of peripheral nerve regeneration and the advantages and disadvantages of artificial nerve conduits and autologous nerve grafts. To overcome the disadvantages, we have developed Bio 3D nerve conduits. We found these Bio 3D nerve conduits to be useful and to have clinical applicability.

Keywords Peripheral nerve · Regeneration · Conduit · 3D · Printer

1 Background

In the surgical treatment of peripheral nerve injury, when the lacerated nerve has no gap, the nerve stumps can be sutured directly. When there is a gap between the nerve stumps, however, the nerve cannot be sutured directly. In such a case, an autologous nerve graft is generally performed. Autologous nerve grafts require the harvesting of healthy nerves and can involve problems such as loss of sensation and pain at the

R. Ikeguchi (✉) · T. Aoyama · H. Yurie · H. Takeuchi · S. Mitsuzawa · M. Ando
S. Ohta · T. Noguchi · S. Matsuda
Department of Orthopaedic Surgery and Rehabilitation Medicine, Kyoto University,
Kyoto, Japan
e-mail: ikeguchi@kuhp.kyoto-u.ac.jp

S. Akieda
Cyfuse Biomedical K.K., Tokyo, Japan

K. Nakayama
Department of Regenerative Medicine and Biomedical Engineering,
Saga University, Saga, Japan

© Springer Nature Switzerland AG 2021
K. Nakayama (ed.), *Kenzan Method for Scaffold-Free Biofabrication*,
https://doi.org/10.1007/978-3-030-58688-1_10

collection sites. It is reported that almost 40% of patients who have undergone nerve grafting suffer from pain at the sacrificed healthy nerve at the donor nerve site [1]. In addition, the donor nerves (e.g., sural nerve, medial antebrachial cutaneous nerve) are sensory nerves in the superficial layer of extremities that are easily harvested so the sources of harvested nerves are thus limited.

Peripheral nerve segments contain Schwann cells and basement membrane, both of which support nerve regeneration. The gold standard method that has thus far been the most promising therapeutic method to regenerate a lacerated nerve with a stump gap is autologous nerve grafting. To solve the problems related to autologous nerve grafts, such as donor site morbidity and limited sources, artificial nerve conduits have been developed. The regeneration capacity of artificial conduits is, however, relatively low compared to autologous nerve grafting because of the lack of cellular elements [2]. Many researchers have attempted to modify the artificial conduits by adding mesenchymal stem cells and Schwann cells [3–5]. Cell death and leakage from the artificial conduits after injection limits the efficacy of these cells. Therefore, we developed a biological, tissue-engineered, and scaffold-free conduit using a Bio 3D printer [6–8].

2 Peripheral Nerve Repair and Regeneration

In the case of peripheral nerve injury caused by a sharp laceration such as a blade cut, direct suturing is possible because there is no gap at the nerve injury site (Fig. 1). Peripheral nerve repair is performed using 10–0 or 9–0 nylon thread under a surgical microscope, and the suture method differs depending on where the needles are passed (Fig. 2). The peripheral nerve has a perineurium that wraps around the nerve axons and an epineurium that surrounds multiple nerve fascicles that are surrounded by the perineurium. Suture techniques are divided into epineurium suture, perineurium suture, and epi-perineurium suture. Epineurium suture involves suturing the

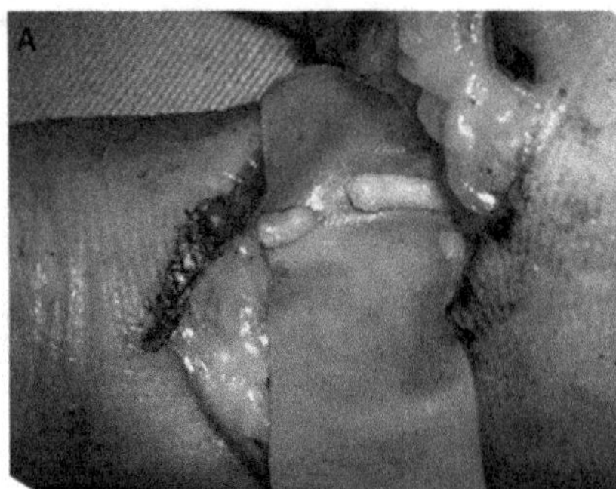
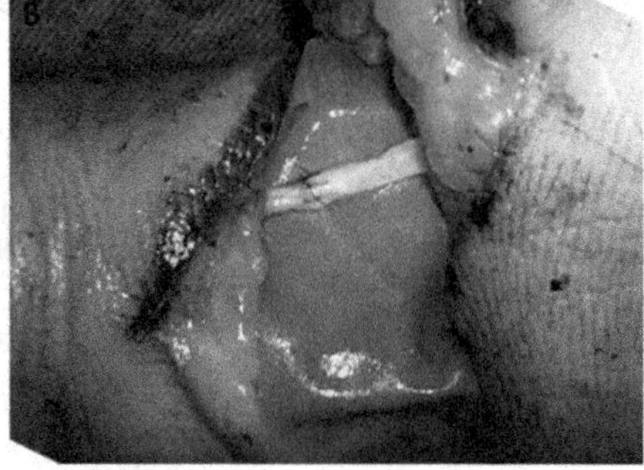

Fig. 1 Case of direct nerve repair. Thirty-nine-year old male suffered a digital nerve laceration by glass. (**a**) Surgical exposure. The digital nerve was lacerated. (**b**) After nerve repair. The nerve was directly sutured

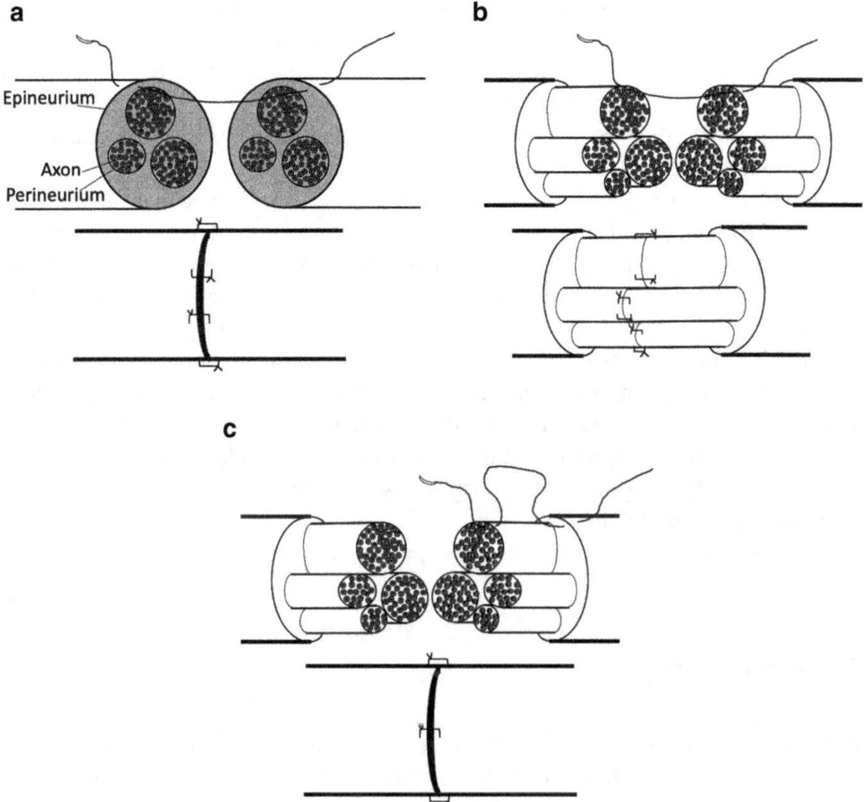

Fig. 2 Type of peripheral nerve suture. (**a**) Epineurium suture. (**b**) Perineurium suture. (**c**) Epi-perineurium suture

epineurium only; perineurium suture involves suturing the perineurium only; and epi-perineurium suture involves suturing both the epineurium and the perineurium. Numerous axons exist inside the perineurium that are important for innervation of target organs. In perineurium suture, suturing needles must be passed through the nerve fascicles, which means that axonal damage cannot be avoided. As such, epineurium suture is recommended when the coaptation of the nerve stump is correct in order to prevent axon damage. In some cases, the nerve fascicles at the nerve stump are inappropriately fitted with the epineurium suture technique. In order to balance the advantages of these two methods, a method called epi-perineural suture was developed. At the surgical field, epineurium suture is sufficient if the nerve fascicles fit well. If the stump fitting is poor, however, perineural suture should be included to improve the coaptation better, and this is called epi-perineural suture.

After nerve repair, the nerve regeneration process occurs. First, the healing mechanism of the injured site occurs and protein synthesis at the stump is promoted. Macrophages gather together at the healing site and capillary vessels are created. In the peripheral nervous system, motor neurons are located in the spinal cord anterior horn and sensory neurons are located in the dorsal root ganglion. The peripheral nerves distal to the lacerated site degenerate after leaving the neuron's cell body, which is referred to as Wallerian degeneration. This is an active process of degeneration in which Schwann cells undergo protein synthesis and are promoted to proliferate at both the proximal and distal stumps and move together with the capillary vessels [9]. Second, the proximal axons branch many regenerated axons (called sprouting) to extend along with the Schwann cells on the created capillary vessels (bands of Büngner) and into the distal nerve stump and the basement membrane reaches the target organ [9]. Third, eventually only axons that reach the target organ, such as the neuromuscular junction and sensory organs, stay alive. The axons that do not reach the target organ are selected out like tree branches, which is called pruning. The migrated Schwann cells surround the elongated axons in a whirling pattern to form myelin, and then the nerve matures. Nerve regeneration requires approximately 1 week to pass through the nerve suture (initial delay). The speed of extension in the distal nerve is 1–2 mm/day (intermediate delay), and in the case of motor nerves, the neuromuscular junction takes several weeks (terminal delay) to regenerate. Put simply, the nerve regeneration rate can be considered to be 1 mm/ day, which is approximately equal to the speed of wound healing.

3   Nerve Grafting

When the peripheral nerve is lacerated by a sharp object such as glass or a blade, direct suturing is possible if surgery takes place within a few days post-injury. When the injury is not sharp due to a machine accident or if the post-injury period is too long even with a sharp cut, a gap is created between the injured nerve stumps, and direct suturing is impossible. In such a case, an autologous nerve graft is required to

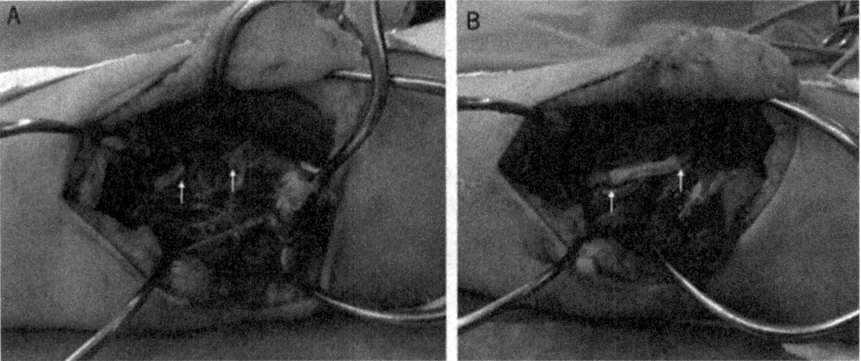

Fig. 3 Case of nerve grafting. Nineteen-year-old female suffered a radial nerve laceration by knife one month ago. (**a**) Surgical exposure. The radial nerve was lacerated with a 5 cm inter-stump gap. Arrows indicate the nerve stumps. (**b**) After nerve grafting. Cable graft was performed with the sural nerve. Arrows indicate the suture sites

bridge the damaged nerve stumps (Fig. 3). The most frequently used donor nerve site is the sensory nerve of the lower leg or forearm. The sural nerve and the medial antebrachial cutaneous nerve are often harvested, because harvesting a motor nerve or important sensory nerves can cause serious damage to motor and sensory function. To reconstruct the small digital nerves, it may be sufficient to use one of the medial forearm nerves, which is determined by the thickness of the damaged nerve and the length of the defect site. To reconstruct the larger nerves, sural nerves should be used in a cable graft fashion, because the defect nerve is thicker and longer (Fig. 3). In these autologous nerve grafts, healthy nerve segments contain Schwann cells and basement membrane, both of which are necessary for the nerve regeneration described above. Autologous nerve grafting is still considered the gold standard because of its prognostic capacity for nerve regeneration [2]. To reconstruct a motor nerve, a sensory nerve segment is also harvested and transplanted. Nerve regeneration extends from the proximal nerve stump, and the transplanted nerve fragment plays a role in bridging the gap and becoming the route for extending axons. If we harvest a motor nerve for motor nerve reconstruction, motor paralysis occurs at the donor site. The sural nerve and medial forearm cutaneous nerve are usually harvested because they can act as a tract for nerve extension and do not cause any problems with motor deficit.

4 Artificial Nerve Conduits

When healthy nerves are harvested in nerve grafting, donor site morbidity such as sensory nerve loss or pain at nerve stumps occurs even though the nerves are sensory nerves, and it may cause a disorder in activities of daily living in some patients that cannot be ignored. Due to the limitations of the nerve source, huge nerve deficits cannot be reconstructed, and some patients cannot undergo surgery of other healthy parts for nerve harvesting in addition to their trauma site. Artificial nerve conduits have been developed as substitutes for autologous nerve transplantation. It was reported in 1982 that the sciatic nerve of rats can be regenerated by a silicon tube [10], and artificial nerves using various materials have subsequently been developed. In the process of nerve regeneration in the silicon tube, the tube is initially filled with an exudate containing nerve growth factor from the nerve stump to form a fibrin matrix, capillaries are created from both nerve stumps along the fibrin matrix, and Schwann cells migrate from both of ends along the capillaries. Finally, axons extend from the proximal stump [11]. The silicon tube that can bridge the nerve defect to regenerate the injured nerve is not absorbable, so absorptive materials have been developed for nerve conduits. The materials are collagen, caprolactone, polyglycolic acid, polyglutamic acid, etc., and some conduits have fibers such as collagen inside them [12]. The design and combination of the inside fibers differ among manufacturing companies.

Nerve regeneration requires various conditions, including scaffold for axonal elongation, growth factors, surrounding blood flow, and supportive cells such as macrophages, fibroblasts, and Schwann cells [13]. Compared to an autologous nerve graft that contains basement membrane and Schwann cells, artificial nerve conduits are composed solely of scaffolds. Therefore, it is theoretically described that the available nerve conduits in the clinical situation have inferior nerve regeneration ability compared to the autologous nerve graft. Artificial nerve conduits are foreign body materials and have a higher risk of infection and less biocompatibility. In a peripheral nerve injury where the surrounding tissue has good blood flow and the implant bed is in good condition without infection or scar tissue, an artificial nerve conduit can be used to bridge the nerve defect within 3 cm of the sensory nerve, especially a digital nerve defect [2]. In this situation, the artificial nerve conduit offers some nerve regeneration and some nerve functional recovery and is acceptable because of the relatively minor sensory nerve. Artificial nerve conduits cannot be used to bridge motor nerve defects and longer sensory defects, however, because of their inferior nerve regeneration capacity and lower expectation of functional recovery. Current clinical situations generally limit the use of the available nerve conduits to acceptable nerve recovery results.

5 Allogeneic Nerve Transplant

As an alternative to artificial nerve conduits, allogeneic nerve transplantation was considered next to autologous nerve grafting [14]. Similar to autologous nerve transplantation, nerve fragments include cell components such as basement membrane and Schwann cells, so that nerve regeneration similar to autologous nerve transplantation can be expected. The function of the transplanted cells requires the use of immunosuppressants, however, which are accompanied by lethal side effects such as infections, malignancies, and renal dysfunctions. The use of immunosuppressants is acceptable for transplantation of vital organs such as the liver, kidney, and heart, but is generally not accepted for many peripheral nerve injuries because they are not life-sustaining organs. In the case of a decellularized allergic nerve segment, the use of an immunosuppressant is not necessary due to its low immunological antigenicity, but nerve regeneration equivalent to autologous nerve transplantation cannot be obtained because it involves no cellular components such as Schwann cells [15]. It is also possible to use immunosuppressive drugs only in the early stages and to gradually reduce the amount of immunosuppressive drugs after nerve regeneration is completed. Some of the Schwann cells that form the myelin sheath consist of graft-derived allogeneic Schwann cells. Therefore, when the immunosuppressive drug is gradually reduced, immunological rejection occurs, and the allogeneic Schwann cells derived from the grafted nerve segment do not function due to rejection, leading to nerve conduction block [16]. In addition, most of the peripheral nerve injuries are semi-emergency operations that take place within a few days after the injuries, and allogeneic nerves may not be able to be prepared before the operations.

6   Nerve Regeneration Using Mesenchymal Stem Cells

Artificial nerve conduits are insufficient for the nerve regeneration described above. As such, other methods using supportive cells have been examined in order to obtain better and longer nerve regeneration. The conditions required for better nerve regeneration include scaffolds growth factors, blood flow in the recipient bed, and cellular components. For scaffolds, conventional commercially available artificial nerves conduits as described above are applicable. For growth factors such as nerve growth factor, better nerve regeneration can be obtained compared to artificial nerve conduits alone. This does not reach the level of autologous nerve grafting, however. Many researchers have attempted to regenerate nerves using supportive cells and have reported better nerve regeneration. Similar to autologous nerve grafts, Schwann cells seem to have some advantages for nerve regeneration, but the isolation and culture of Schwann cells is difficult and the amount of the transplantation is limited after collecting autologous Schwann cells. Therefore, some researchers considered applying mesenchymal stem cells (MSCs). Among MSCs, bone marrow-derived cells (BMSCs) derived from bone marrow and adipose tissue-derived cells (ASCs) derived from fat are widely used [17]. Both BMSCs and ASCs can be collected and separated and cultured relatively easily and are employed in various fields of regenerative medicine as well as for easy peripheral nerve regeneration. It is known that BMSCs differentiate into Schwann-like cells with the Schwann cell phenotype in vitro and the mechanism is becoming clear [18, 19]. It has been confirmed in in vivo studies that BMSCs transplanted into peripheral nerve defects differentiate into Schwann-like cells and promote axonal regeneration and myelination, leading to better nerve regeneration [4, 5]. It has been reported that ASCs promote nerve regeneration similar to BMSCs [20], and there are also some reports of dental pulp-derived MSCs and peripheral blood-derived MSCs [21, 22]. The survival rate of the transplanted cells after 2 weeks has been reported to be 5.8% [23]. Various methods of applying cells have been tried, such as simple injection and mixing with fibrin glue. Supportive cells are not, however, applied effectively for nerve defect sites and the nerve regeneration of artificial nerve conduits with supportive cells is not superior to that of autologous nerve grafts. Further research is needed to develop cell-based nerve conduits.

7   Bio 3D Nerve Conduit

Bio 3D printing technology, which constructs living tissue three-dimensionally, is already being applied in many regenerative medicine fields, such as blood vessels, trachea, and cartilage [24–26]. Nakayama et al. developed technology that uses the Kenzan method to construct a complex three-dimensional structure only from cells (without using scaffolds). Especially in the field of blood vessels, spheroids are prepared by aggregating cells using three types of cells (vascular endothelial cells, smooth muscle cells, and fibroblasts), and based on 3D data, blood vessel structures are artificially created from the spheroids using the Bio 3D printer Regenova® [24]. The advantages of the Bio 3D printer are that the structure (size, shape, and length) can be freely designed according to the clinical setting using computer-aided design, and the created tissue does not contain foreign materials, which may induce foreign body reactions, infection, or allergy.

As described above, the challenges for better nerve regeneration of artificial nerve conduits were addressed by adding Schwann cells or mesenchymal cells as support-ing cells. The survival of administered cells in the nerve conduits is relatively low, however [23]. Therefore, we focused on a completely biological, tissue-engineered, scaffold-free conduit [8]. It is creating a nerve conduit consisting solely of these sup-porting cells (Bio 3D nerve conduit) using Bio 3D printing technology. We devel-oped Bio 3D nerve conduits from human normal dermal fibroblasts. A Bio 3D nerve conduit is strong enough for surgical handling and microsurgical suture (Fig. 4). We evaluated the nerve regeneration capacity of the Bio 3D nerve conduit using a sciatic nerve defect model of immune-deficiency rats. We created a nerve gap at the mid-thigh level by transecting the right sciatic nerve and bridged the gap with Bio 3D nerve conduits (Fig. 5). Kinematic, electrophysiological, and morphological assess-ments were conducted to examine nerve regeneration 8 weeks post-surgery. Kinematic analysis revealed significantly higher movement of the metatarsophalan-geal joint in the Bio 3D nerve conduit group than that in the control group.

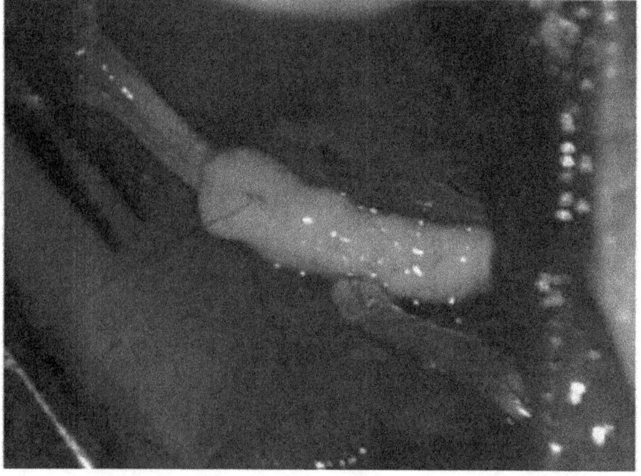

Fig. 4 Picture of a Bio 3D nerve conduit during surgery under the operating microscope. The conduit is strong enough to pass the thread

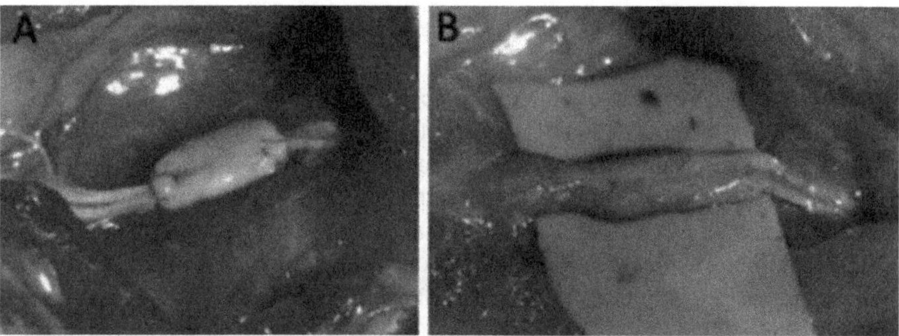

Fig. 5 Nerve regeneration by a Bio 3D nerve conduit. (**a**) A rat sciatic nerve gap is bridged with a Bio 3D nerve conduit. (**b**) Eight weeks after surgery. Nerve regeneration is evident macroscopically

Electrophysiological studies revealed significantly higher action potential in the pedal adductor muscle in the Bio 3D nerve conduit group than in the control group. Morphological studies revealed many well-myelinated axons in the Bio 3D nerve conduit group. Muscle weight analysis revealed that the tibialis anterior muscle was significantly larger in the Bio 3D nerve conduit group than in the control group. Thus, we confirmed that the Bio 3D nerve conduits composed of fibroblast cells promote nerve regeneration. The process of nerve regeneration requires not only Schwann cells but also macrophages and fibroblasts [13]. We utilized fibroblasts because they are easy to culture and proliferate in vitro and exhibit promising mechanical strength. In addition, the possibility of the differentiation of fibroblasts into Schwann cells and mesenchymal cells has been reported, and the nerve regeneration process requires fibroblasts [27]. The advantage of Bio 3D printing is that the structure is designed based on data created on a computer, and the shape and length of the nerve conduit can be freely changed according to the nerve defects. In addition, the Bio 3D nerve conduit is pure biological material and can deliver viable supportive cells. A Bio 3D printer can be utilized to construct multilayer Bio 3D nerve conduits composed of multicellular spheroids from BMSCs or any other cells to improve the seeding efficacy and viability of the cells.

There are some difficulties for clinical application, however. It takes a few weeks to collect and culture cells in a clinical setting and to create a structure. For emergency surgery, we should use allogeneic less immunoreactive cells such as HLA-matched iPS cells to prepare the Bio 3D nerve conduit before the trauma. After transplantation, joint movement may collapse the conduits. As such, the nerve conduit requires a certain strength and we developed stronger conduit and evaluated their strength precisely. We will also keep researching the usefulness for longer nerve deficits. We reported good peripheral nerve regeneration using Bio 3D nerve conduits in 5 mm and 10 mm nerve defects [6–8]. Evaluation of the usefulness of Bio 3D conduits in long nerve deficits is required. Further research is needed for future clinical applications.

8　Conclusion

We described the surgical treatments for peripheral nerve injuries and the basic research that has been conducted. At present, in cases where there is a gap in nerve stumps, the most promising treatment for nerve regeneration is autologous nerve transplantation. Studies on nerve regeneration using Bio 3D nerve conduits have been conducted, however, and it is known that they promote nerve regeneration better than clinically available artificial nerve conduits. In the future, a new therapeutic method may arise if the usefulness and applicability of Bio 3D nerve conduits are further advanced.

References

1. Staniforth P, Fisher TR (1978) The effects of sural nerve excision in autogenous nerve grafting. Hand 10(2):187–190
2. Griffin JW, Hogan MV, Chhabra AB et al (2013) Peripheral nerve repair and reconstruction. J Bone Joint Surg Am 95(23):2144–2151
3. Jesuraj NJ, Santosa KB, Newton P et al (2011) A systematic evaluation of Schwann cell injection into acellular cold-preserved nerve grafts. J Neurosci Methods 197:209–215
4. Yamakawa T, Kakinoki R, Ikeguchi R et al (2007) Nerve regeneration promoted in a tube with vascularity containing bone marrow-derived cells. Cell Transplant 16:811–822
5. Kaizawa Y, Kakinoki R, Ikeguchi R et al (2017) A nerve conduit containing a vascular bundle and implanted with bone marrow stromal cells and decellularized allogenic nerve matrix. Cell Transplant 26(2):215–228
6. Takeuchi H, Ikeguchi R, Aoyama T et al (2019) A scaffold-free Bio 3D nerve conduit for repair of a 10-mm peripheral nerve defect in the rats. Microsurgery. https://doi.org/10.1055/s00393400529
7. Mitsuzawa S, Ikeguchi R, Aoyama T et al (2019) The efficacy of a scaffold-free Bio 3D conduit developed from autologous dermal fibroblasts on peripheral nerve regeneration in a canine ulnar nerve injury model: a preclinical proof-of-concept study. Cell Transplant 28(9–10):1231–1241
8. Yurie H, Ikeguchi R, Aoyama T et al (2017) The efficacy of a scaffold-free Bio 3D conduit developed from human fibroblasts on peripheral nerve regeneration in a rat sciatic nerve model. PLoS One. https://doi.org/10.1371/e0171448
9. Jessen KR, Mirsky R, Lloyd AC (2015) Schwann cells: development and role in nerve repair. Cold Spring Harb Perspect Biol. https://doi.org/10.1101/a020487
10. Lundborg G, Dahlin LB, Danielsen N et al (1982) Nerve regeneration in silicone chambers: influence of gap length and of distal stump components. Exp Neurol 76(2):361–375
11. Williams LR, Powell HC, Lundborg G et al (1984) Competence of nerve tissue as distal insert promoting nerve regeneration in a silicone chamber. Brain Res 293(2):201–211
12. Houshyar S, Bhattacharyya A, Shanks R et al (2019) Peripheral nerve conduit: materials and structures. ACS Chem Neurosci 10(8):3349–3365
13. Pan D, Mackinnon SE, Wood MD (2019) Advances in the repair of segmental nerve injuries and trends in reconstruction. Muscle Nerve. https://doi.org/10.1002/mus26797
14. Bain JR, Mackinnon SE, Hudson AR et al (1988) The peripheral nerve allograft: an assessment of regeneration across nerve allografts in rats immunosuppressed with cyclosporin a. Plast Reconstr Surg 82(6):1052–1064

15. Han LW, Xu G, Guo MY et al (2019) Comparison of SB-SDS and other decellularization methods for the acellular nerve graft: biological evaluation and nerve repair in vitro and in vivo. Synapse. https://doi.org/10.1002/e22143

16. Midha R, Mackinnon SE, Evans PJ et al (1993) Comparison of regeneration across nerve allografts with temporary or continuous cyclosporin A immunosuppression. J Neurosurg 78(1):90–100

17. Kubiak CA, Grochmal J, Kung TA et al (2019) Stem-cell-based therapies to enhance peripheral nerve regeneration. Muscle Nerve. https://doi.org/10.1002/mus26760

18. Wakao S, Matsuse D, Dezawa M (2015) Mesenchymal stem cells as a source of Schwann cells: their anticipated use in peripheral nerve regeneration. Cells Tissues Organs 200(1):31–41

19. Sharma AD, Wiederin J, Uz M et al (2017) Proteomic analysis of mesenchymal to Schwann cell transdifferentiation. J Proteome 165:93–101

20. Kingham PJ, Kolar MK, Novikova LN et al (2014) Stimulating the neurotrophic and angiogenic properties of human adipose-derived stem cells enhances nerve repair. Stem Cells Dev 23(7):741–754

21. Kolar MK, Itte VN, Kingham PJ et al (2017) The neurotrophic effects of different human dental mesenchymal stem cells. Sci Rep. https://doi.org/10.1038/s41598017129691

22. Pan M, Wang X, Chen Y et al (2017) Tissue engineering with peripheral blood-derived mesenchymal stem cells promotes the regeneration of injured peripheral nerves. Exp Neurol 292:92–101

23. Walsh SK, Kumar R, Grochmal JK et al (2012) Fate of stem cell transplants in peripheral nerves. Stem Cell Res 8(2):226–238

24. Itoh H, Nakayama K, Noguchi R et al (2015) Scaffold-free tubular tissues created by a Bio-3D printer undergo remodeling and endothelialization when implanted in rat aortae. PLoS One 2015. https://doi.org/10.1371/e0136681

25. Dikina AD, Strobel HA, Lai BP et al (2015) Engineered cartilaginous tubes for tracheal tissue replacement via self-assembly and fusion of human mesenchymal stem cell constructs. Biomaterials 52:452–462

26. Ishihara K, Nakayama K, Akieda S et al (2014) Simultaneous regeneration of full-thickness cartilage and subchondral bone defects in vivo using a three-dimensional scaffold-free autologous construct derived from high-density bone marrow-derived mesenchymal stem cells. J Orthop Surg Res 9:98

27. Thoma EC, Merkl C, Heckel T et al (2014) Chemical conversion of human fibroblasts into functional Schwann cells. Stem Cell Rep 3(4):539–547

Regeneration of the Diaphragm

Yusuke Yanagi, Xiu-Ying Zhang, Kouji Nagata, and Tomoaki Taguchi

Abstract The diaphragm is an essential mammalian skeletal muscle for respiration, comprising muscle, muscle connective tissue, tendon, nerves, and vasculature. Failed diaphragmatic development causes diaphragmatic defects in neonates. Neonates with congenital diaphragmatic hernia often require surgical defect closure with a patch. However, the clinical efficacy of patches is limited by complications associated with residual foreign material and by hernia recurrence. Engineered diaphragmatic repair is an emerging area of regenerative medicine in pediatric surgery. We fabricated scaffold-free cellular patches composed of human cells using the "Kenzan" method. The engineered cellular patches had a high elasticity and strength and were transplanted into rats with surgically created diaphragmatic defects. Diaphragmatic hernia was not observed in the animals, and no mortality was recorded during rat growth. Histology revealed regeneration of the muscle structure, neovascularization, and neuronal networks within the reconstructed diaphragms. Our results show that our newly created cellular patches are an extremely safe and effective therapeutic strategy for repairing diaphragmatic defects.

Keywords Tissue engineering · 3D bioprinter · Scaffold-free cellular patch · Congenital diaphragmatic hernia · Skeletal muscle regeneration

Y. Yanagi (✉) · X.-Y. Zhang · K. Nagata · T. Taguchi
Department of Pediatric Surgery, Reproductive and Developmental Medicine, Kyushu University Graduate School of Medical Sciences, Higashi-ku, Fukuoka, Japan
e-mail: y-yanag@pedsurg.med.kyushu-u.ac.jp

© Springer Nature Switzerland AG 2021
K. Nakayama (ed.), *Kenzan Method for Scaffold-Free Biofabrication*,
https://doi.org/10.1007/978-3-030-58688-1_11

Abbreviations

ACM	acellular collagen matrix
CDH	congenital diaphragmatic hernia
HUVEC	human umbilical vein endothelial cell
MCS	multicellular spheroid
NHDF	normal human dermal fibroblast
PPFs	pleuroperitoneal folds
SIS	small intestinal submucosa

1 Introduction

The diaphragm is an important mammalian skeletal muscle. It is essential for respiration and for separating the thoracic and abdominal cavities. Diaphragm contraction drives inspiration [1]. Development of the diaphragm requires the coordinated development of muscle, muscle connective tissue, tendon, nerves, and vasculature that derive from different embryonic sources [2]. The failure of diaphragmatic development causes diaphragmatic defects in neonates that often result in the lethal birth defect congenital diaphragmatic hernia (CDH).

It was recently reported that pleuroperitoneal folds (PPFs), which are transient embryonic structures, are the source of the diaphragm's muscle connective tissue and regulate muscle development. The migration and expansion of PPF-derived fibroblasts precede muscle morphogenesis, formation of the vascular network, and growth of the phrenic nerve. Gata4 mosaic mutations in PPF-derived muscle connective tissue fibroblasts result in the development of localized amuscular regions that are biomechanically weaker and more compliant, leading to CDH. Expansion and morphogenesis of the fibroblastic connective tissue are therefore critical for diaphragm formation [3].

2 Congenital Diaphragmatic Hernia

Congenital defect of the diaphragm causes CDH. This defect occurs in 1/3000 total births [4]. In CDH, the abdominal organs herniate into the thoracic cavity and impede the lung development, leading to neonatal ventilatory insufficiency and pulmonary hypertension. The associated lung hypoplasia is the cause of the 50% neonatal mortality and long-term morbidity associated with CDH [5].

Although many candidates of CDH-associated genes have been identified, the majority of CDH cases are isolated [6, 7]. Several factors have been suggested as the etiology of CDH, including vitamin A deficiency, chromosomal abnormality, and single-gene mutation [5]. However, the etiology of CDH is incompletely understood.

Therapy for CDH involves closure of the diaphragmatic defect. The size of the defect varies from a small deficiency of the posterior muscular rim to a complete absence of the diaphragm [8]. Small defects can be repaired by primary closure using the remnant diaphragm. In contrast, large defects often require surgical closure with synthetic patches [9, 10]. Patch repair for large diaphragmatic defects was required in around 50% of cases [8]. Synthetic durable sheets, like Gore-Tex (polytetrafluorethylene [PTFE], W.L. Gore & Associates, Flagstaff, AZ, USA), can be considered the gold standard material for CDH repair. However, patch repair of CDH is associated with a high rate of hernia recurrence and can result in thoracic deformity, as the synthetic material does not grow with the infant [11, 12]. In addition, the use of prosthetic materials, which are foreign substances, may lead to granulation, allergic reaction, infection, and small bowel obstruction [12, 13]. Thus, there is a great need for large patches that can grow with the patients and for patches made without foreign material.

3 Tissue Engineering of the Diaphragm

To overcome the complications associated with prosthetic patches, tissue engineering approaches have been explored. The ideal scaffold for the diaphragm is non-toxic, durable, elastic, and biodegradable. Many kinds of engineered patches, including natural and biosynthetic materials, have been evaluated for their suitability for closure of diaphragmatic defects [14]. Several acellular collagen matrix (ACM) patches are available on a commercial basis, and clinical evidence has shown that such patches induce greater collagen deposition and less of an inflammatory response [15]. Surgisis (Cook Medical, Strombeek-Bever, Belgium) is a porcine non-cross-linked collagen sheet derived from small intestinal submucosa (SIS). A comparative study of the recurrent hernia ratio after patch repair of CDH between Surgisis and prosthetic patches showed no advantage of Surgisis over other patches. The recurrent herniation was considered to be related to the rapid degradation of the sheet [16–19]. Matrigel, which is a cross-linked ACM, is made of highly purified porcine type I collagen and contains low amounts of other natural fiber-forming proteins, like elastin and collagen type III. However, Matrigel was not proven to be a viable option for CDH repair, since it has lower compliance than native tissue [20].

Hybrid scaffold patches have also been explored in order to capitalize on the strong points of different materials. A poly-lactic-co-glycolic acid (PLGA) mesh-collagen sponge hybrid scaffold was implanted into rats in which diaphragmatic defects had been created surgically. As a result, autologous fibrous tissue with vascularization was generated at the graft site. However, no muscular tissue was detected [21]. Aligned electrospun poly(ε-caprolactone) (PCL)/collagen hybrid scaffolds showed no evidence of herniation or retraction up to 6 months after implantation. Histological evaluations revealed ingrowth of muscle tissue into the scaffolds. However, neuronal network formation in the reconstructed diaphragm was not confirmed after transplantation [22].

The combination of a scaffold and cells to create a tissue engineered diaphragm is a novel approach, as it may combine a high regenerative capacity and rapid growth. An amniocyte-based engineered tendon was a three-layer patch seeded with autologous amniotic mesenchymal stem cells. The scaffold consisted of trilayered composites made of 70% type I collagen hydrogel solution placed in between acellular human dermis and SIS. While this approach resulted in better outcomes than an acellular bioprosthetic patch, skeletal muscle fibers could not be identified morphologically [23]. The decellularized diaphragmatic tissue reseeded with bone marrow mesenchymal stromal cells was able to replace 80% of the left hemidiaphragm of rats. It facilitated the regeneration of the functional diaphragm, and a histological evaluation showed similarity to native tissue. However, donation of the diaphragmatic tissue for decellularized scaffold limits this approach [24].

4 Regeneration of the Diaphragm with a Bio-3D Cellular Patch

The usage of exogenous materials in scaffolds is associated with a number of problems, including infection, immune reactions, and the degradation of materials and biofilms [25]. We therefore attempted to fabricate scaffold-free patches composed of human cells using the "Kenzan" method [26].

We first created a sheet-shaped scaffold-free fibroblast patch. The fibroblast spheroids were placed into the needle array in the shape of a single-layer sheet. As the tensile strength of the patch was not adequate for transplantation, to improve it, we adopted the following three strategies: create a tubular shape and perform culture using a circulation system, perform co-culture of human umbilical vein endothelial cells (HUVECs) with normal human dermal fibroblasts (NHDFs), and prolong the culture duration. The multicellular spheroid (MCS) composed of HUVECs and NHDFs showed approximately 98% cell viability.

According to a 3D structure predesigned by a computer, we bio-printed MCSs to fabricate a tube-shaped tissue (Fig. 1a). The cytotoxicity assay showed that the needles damaged only a slight number of cells. After 7 weeks of culture, the tubular tissue had formed, measuring approximately 8 mm in length with a wall thickness of approximately 1 mm (Fig. 1a). During tissue organization, the cell fusion induced about 60% shrinkage of the tissue. The tubular tissue had a smooth surface and a perfectly elastic structure. A histological analysis revealed the presence of microvessels and production of the extracellular matrix throughout the tissue (Fig. 2a–f). We confirmed the self-production of collagen I by the constituent cells (Fig. 2e, f). The diaphragm is a flat sheet, so the created tubular tissue was cut into rectangles to form a sheet shape (Fig. 1a). The tensile strength of the tissue cultured under optimized conditions (tubular shape with circulation culture, composed of 90% NHDFs and 10% HUVECs, 7-week culture duration) was equivalent to that of the rat diaphragm.

We repaired the surgically created diaphragmatic defects in rats using the cellular patches (Fig. 1b, c). The treated rats gained body weight at a rate similar to that of normal rats. Furthermore, in the cellular patch repair group, diaphragmatic hernia was not observed, nor were any mortalities recorded at any time. The longest survival was up to 710 days after operation. These results suggest that the cellular patches were effective for repairing diaphragmatic defects during rat growth. Upon a gross examination at the time of sacrifice, the implant cellular patch was successfully integrated with the host diaphragm, with clearly visible areas of vascularization. No herniation and no rib cage deformation were present. Slight adhesions between the stomach and the abdominal side of the patch were observed. However, no adhesion with the lung was present.

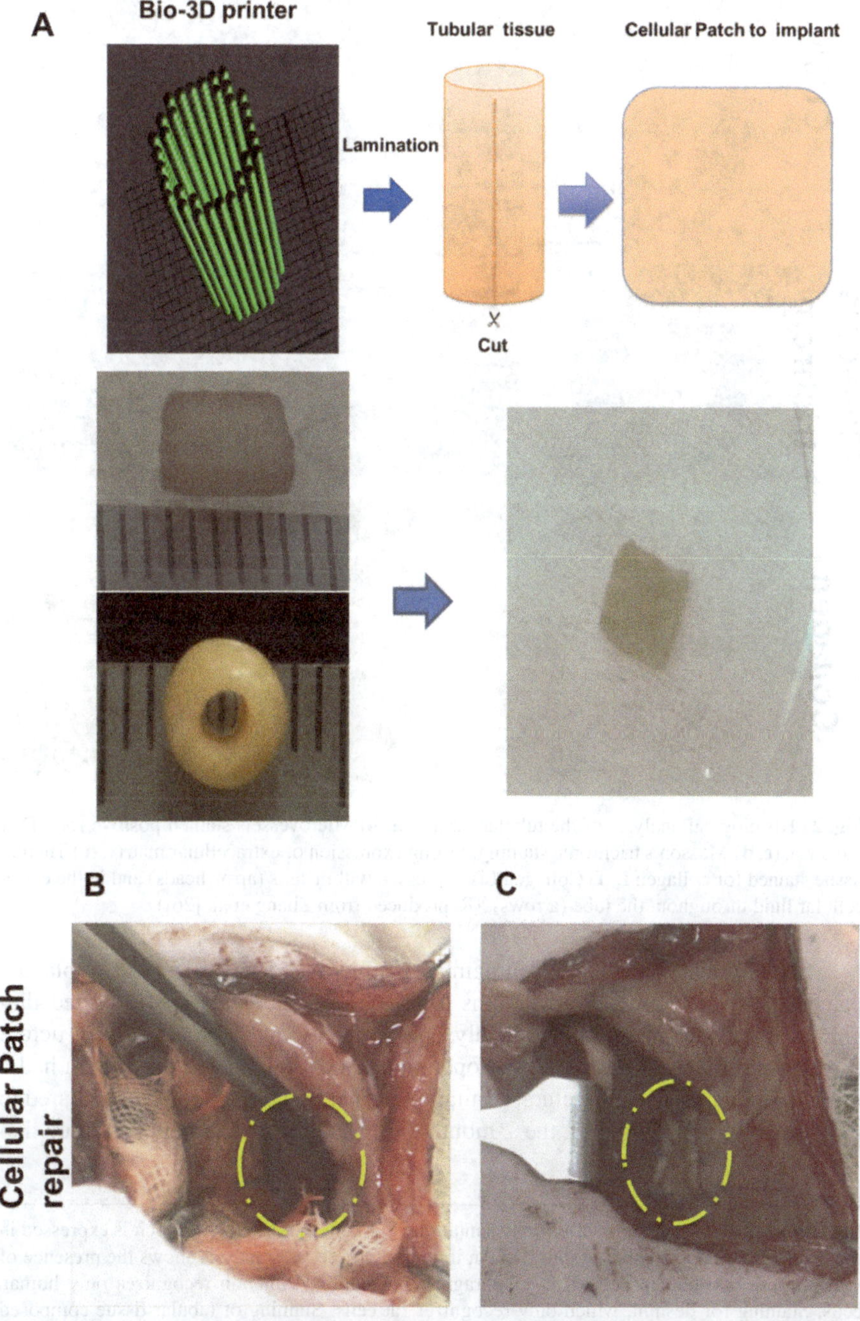

Fig. 1 The cellular patch for diaphragmatic repair constructed by the "Kenzan" method. (**a**) A simple schematic illustration of the method for creating a patch containing only cells. First, tubular tissue is created; the resulting tissue is then cut into a patch for implantation. (**b**) A diaphragmatic defect approximately 12 × 10 mm in size was made in a rat. The yellow circle shows the defect in the diaphragm. (**c**) A cellular patch was implanted in the rat diaphragm. (Reproduced from Zhang et al. [26])

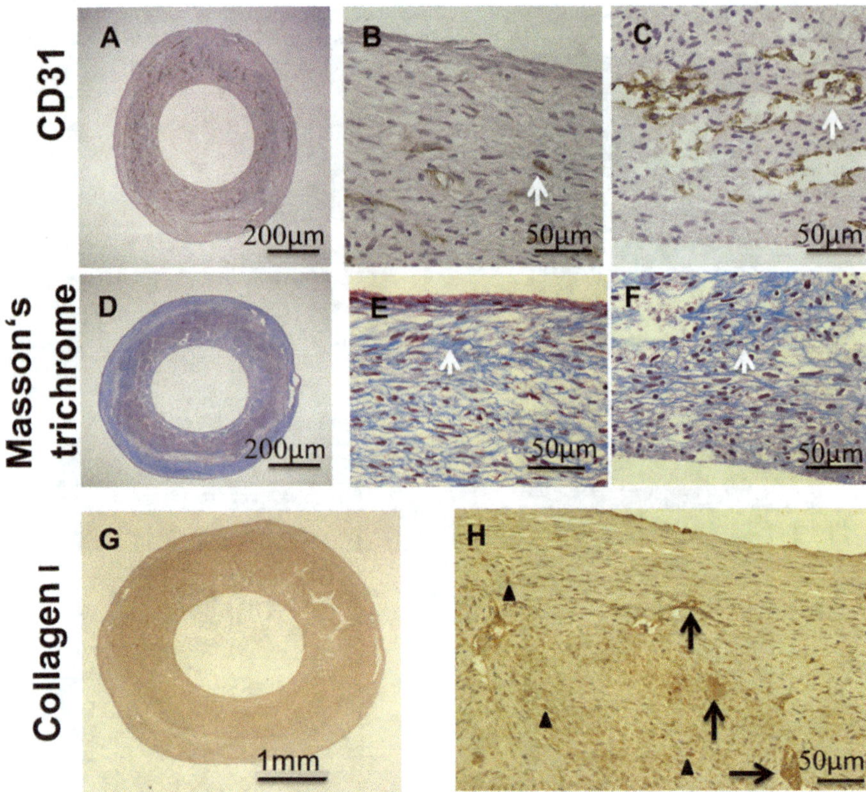

Fig. 2 Histological analysis of the tubular tissue. (**a, b**) Microvessels stained positive for CD31 (arrows). (**c, d**) Masson's trichrome staining. Strong expression of extracellular matrix. (**e**) Tubular tissue stained for collagen I. (**f**) Collagen I is expressed within cells (arrowheads) and in the extracellular fluid throughout the tube (arrows). (Reproduced from Zhang et al. [26])

After removal from the diaphragm, a histological analysis of the implanted patches was performed. Four months after the operation, the reconstructed diaphragm showed clear morphologically normal nuclei and a muscle-like structure (Fig. 3a–c). Seven months after the operation, the reconstructed diaphragm had a more normal muscle architecture, similar to normal tissue (Fig. 3d–f). Compared to the 4-month group (Fig. 3b), the 7-month group showed a markedly increased dis-

Fig. 3 (continued) diaphragm after transplantation. (**j**) Staining for MAP 2, which is expressed in the cell bodies and axons of neurons. (**k**) An immunofluorescence analysis shows the presence of vimentin-positive human cells after diaphragm regeneration. Vimentin recognizes only human cells. Staining for desmin, which only recognizes rat cells. Staining of tubular tissue composed only of human cells shows vimentin-positive cells. Staining for desmin is negative (median panel). In normal rat diaphragm muscle, desmin-positive rat cells are observed, while vimentin-positive human cells are not present (upper panel). In the reconstructed diaphragm 7 months after transplantation, vimentin-positive human cells are still clearly visible. These cells are localized along the regenerated striated muscle, which is stained with desmin, and human cells were also localized in the regenerating muscle (lower panel). (Reproduced from Zhang et al. [26])

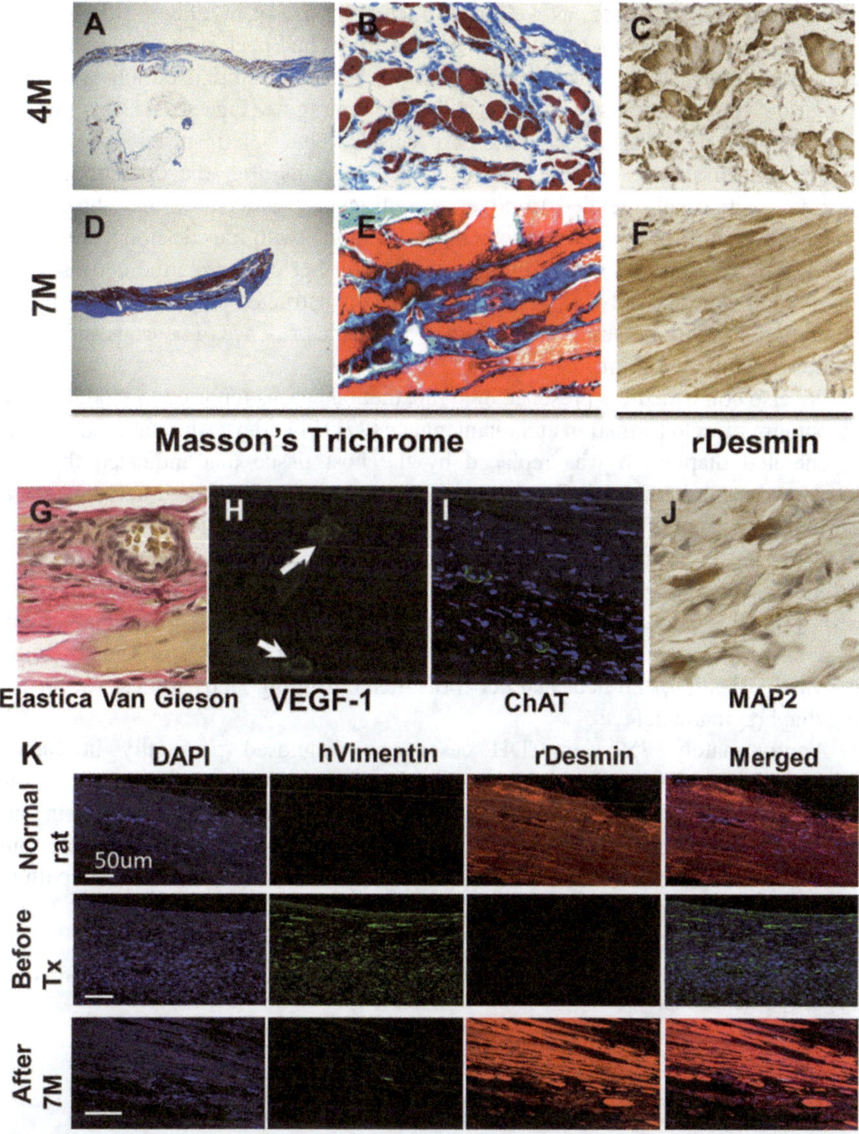

Fig. 3 A histological analysis of the reconstructed diaphragm after transplantation. (a–e) Masson's trichrome staining. The grafts in the transplanted area after 4 months (a) and 7 months (d). Magnified images show collagen fibers around the muscle after 4 months (b) and 7 months (d). Seven months after transplantation, the shape and size distribution of the myofibers is close to normal (d). (c, f) Desmin staining. At 4 months after the operation, new muscle cells developed in the sheet-shaped structure (c). At 7 months after the operation, the shape and size distribution of the muscle tissue was close to normal (f). (g–j) The presence of vasculature and neurons in reconstructed diaphragm after transplantation. (g) EVG staining shows many elastic fibers with a vessel wall-like structure around the muscle in the reconstructed diaphragm. (h) Staining for VEGF-1, which is specifically expressed in most vascular endothelial cells. VEGF-1 IR-positive cells (arrows) are observed in the reconstructed diaphragm after transplantation. (i) Staining for ChAT, which is a motor neuron marker. ChAT IR-positive cells are visible (green) in the reconstructed

tribution of collagen fibers as well as an increased size of myofibers (Fig. 3e). To further investigate whether or not the muscle fibers had been properly formed, we evaluated the structural muscle protein desmin. Desmin immunostaining was positive at the implantation site in reconstructed diaphragms (Fig. 3c, f). At 4 months after the operation, the size and number of muscle fibers were insufficient compared to the control striated muscle. However, at 7 months after the operation, these desmin-positive cells tended to cover a significantly wider area and showed an increased volume compared to the cells at 4 months, with muscle fibers similar to those in normal rats (Fig. 3f). Immunostaining further confirmed the neovascularization and neuronal network formation in the reconstructed diaphragm after transplantation (Fig. 3g–j). Additional testing revealed the presence of neurons and axons in the reconstructed graft.

We also confirmed the presence of sympathetic neurons reported to modulate the diaphragm muscle formation and maintenance [27] (Fig. 3i, j). Although most of the regenerated diaphragm was replaced by the host tissue that indicated that the implanted patch induced the host tissues such as muscle, vessel, and nerve, we confirmed that human cells had integrated into the structure of the regenerated diaphragm to a substantial extent (Fig. 3k). This finding showed that fibroblast patches could promote the recruitment and differentiation of endogenous myogenic progenitor cells [28, 29]. This process may recapitulate the development of the diaphragm. Conclusively, our cellular patch induced the regeneration of the muscle structure, neovascularization, and neuronal networks, demonstrating its usefulness for repairing diaphragmatic defects.

Approximately 75% of CDH cases are diagnosed prenatally in Japan. Diaphragmatic repair should be performed early after the stabilization of the patient's vital condition [30]. Cellular patches should therefore be fabricated in parallel to gestation so as to be promptly available after stabilization. Therefore, our next step should involve exploring appropriate cell sources, reducing the culture duration, and fabricating human-sized cellular patches.

Acknowledgments The authors thank Mr. Brian Quinn for his help in the preparation of the English version of this article.

References

1. Perry SF et al (2010) The evolutionary origin of the mammalian diaphragm. Respir Physiol Neurobiol 171:1–16
2. Sefton EM et al (2018) Developmental origin and morphogenesis of the diaphragm, an essential mammalian muscle. Dev Biol 440:64–73
3. Merrell A et al (2015) Muscle connective tissue controls development of the diaphragm and is a source of congenital diaphragmatic hernias. Nat Genet 47:496–505
4. Torfs CP et al (1992) A population-based study of congenital diaphragmatic hernia. Teratology 46:555–565
5. Pober BR (2007) Overview of epidemiology, genetics, birth defects, and chromosome abnormalities associated with CDH. Am J Med Genet C Semin Med Genet 145c:158–171

6. Holder AM et al (2007) Genetic factors in congenital diaphragmatic hernia. Am J Hum Genet 80:825–845
7. Veenma DC et al (2012) Developmental and genetic aspects of congenital diaphragmatic hernia. Pediatr Pulmonol 47:534–545
8. Congenital Diaphragmatic Hernia Study Group, Lally KP et al (2007) Defect size determines survival in infants with congenital diaphragmatic hernia. Pediatrics 120:651–657
9. Grethel EL et al (2006) Prosthetic patches for congenital diaphragmatic hernia repair: Surgisis vs Gore-Tex. J Pediatr Surg 41:29–33
10. Tovar JA (2012) Congenital diaphragmatic hernia. Orphanet J Rare Dis 7
11. Tsao K, Lally KP (2011) Surgical management of the newborn with congenital diaphragmatic hernia. Fetal Diagn Ther 29:46–54
12. Gasior AC, St Peter SD (2012) A review of patch options in the repair of congenital diaphragm defects. Pediatr Surg Int 28:327–333
13. Riehle KJ et al (2007) Low recurrence rate after Gore-Tex/Marlex composite patch repair for posterolateral congenital diaphragmatic hernia. J Pediatr Surg 42:1841–1844
14. Fauza DO (2014) Tissue engineering in congenital diaphragmatic hernia. Semi Pediatr Surg 23:135–140
15. Deprest J et al (2006) The biology behind fascial defects and the use of implants in pelvic organ prolapse repair. Int Urogynecol J Pelvic Floor Dysfunct 17:S16–S25
16. Dalla VL et al (1999) Evaluation of small intestine submucosa and acellular dermis as diaphragmatic prostheses. J Pediatr Surg 34:167–171
17. Sandoval JA et al (2006) The whole truth: comparative analysis of diaphragmatic hernia repair using 4-ply vs 8-ply small intestinal submucosa in a growing animal model. J Pediatr Surg 41:518–523
18. Rodrigo LP et al (2012) What is the best prosthetic material for patch repair of congenital diaphragmatic hernia? Comparison and meta-analysis of porcine small intestinal submucosa and polytetrafluoroethylene. J Pediatr Surg 47:1496–1500
19. Grethel EJ et al (2006) Prosthetic patches for congenital diaphragmatic hernia repair: Surgisis vs Gore-Tex. J Pediatr Surg 41:29–33
20. Mayer S et al (2015) Diaphragm repair with a novel cross-linked collagen biomaterial in a growing rabbit model. PLoS One 10:e0132021
21. Urita Y et al (2008) Evaluation of diaphragmatic hernia repair using PLGA mesh–collagen sponge hybrid scaffold: an experimental study in a rat model. Pediatr Surg Int 24:1041–1045
22. Zhao W et al (2013) Diaphragmatic muscle reconstruction with an aligned electrospun poly(ε-caprolactone)/collagen hybrid scaffold. Biomaterials 34:8235–8240
23. Turner CG et al (2011) Preclinical regulatory validation of an engineered diaphragmatic tendon made with amniotic mesenchymal stem cells. J Pediatr Surg 46:57–61
24. Gubareva EA et al (2016) Orthotopic transplantation of a tissue engineered diaphragm in rats. Biomaterials 77:320–335
25. Nakayama K (2013) Biofabrication. In: Forgacs J (ed) Micro- and nano-fabrication, printing, patterning, and assemblies. Ch.1. Elsevier, Amsterdam, pp 1–16
26. Zhang XY et al (2018) Regeneration of diaphragm with bio-3D cellular patch. Biomaterials 167:1–14
27. Lustrino D et al (2017) Sympathetic innervation controls homeostasis of neuromuscular junctions in health and disease. Proc Natl Acad Sci 114:e5277–e5277
28. Sicari BM et al (2014) Tissue engineering and regenerative medicine approaches to enhance the functional response to skeletal muscle injury. Anat Rec 297:51–64
29. Vining KH, Mooney DJ (2017) Mechanical forces direct stem cell behaviour in development and regeneration. Nat Rev Mol Cell Biol 19:728–742
30. Nagata K et al (2013) The current profile and outcome of congenital diaphragmatic hernia: a nationwide survey in Japan. J Pediatr Surg 48:738–744

Bio-3D Printed Organs as Drug Testing Tools

Kenichi Arai and Koichi Nakayama

Abstract Evaluations of side effects and therapeutic efficacy are key steps in drug development. This process is time-consuming and costly and is often discontinued owing to differences in the drug response between animal models and humans, necessitating alternative approaches to animal models. One promising alternative for evaluations of therapeutic efficacy and adverse reactions is human induced pluripotent stem cells (iPS cells). Human iPS cells can differentiate into various cell types. Furthermore, using iPS cells derived from patients, useful human disease models can be established for the development of drugs against a wide range of diseases. Although cells are cultured on a culture dish, spheroids and organoids, which are aggregates of cells with tissue-specific functions, develop to mimic the three-dimensional (3D) cell environment. Bio-3D printing has been developed to generate scaffold-free constructs by using spheroids according to a desired design. The combination of organoids and bio-3D printing is expected to enable the fabrication of constructs with tissue-specific functions for drug testing. In this chapter, we discuss the current status of drug development and tissue engineering techniques as well as the fabrication of spheroids or organoids and applications for drug evaluation. Finally, we discuss the existing challenges and future prospects.

Keywords Bio-3D printer · Drug response · Side effect · Spheroid · Organoid

K. Arai (✉) · K. Nakayama
Center for Regenerative Medicine Research, Faculty of Medicine, Saga University, Saga, Japan
e-mail: kenarai@med.u-toyama.ac.jp; info@nakayama-labs.com

© Springer Nature Switzerland AG 2021
K. Nakayama (ed.), *Kenzan Method for Scaffold-Free Biofabrication*,
https://doi.org/10.1007/978-3-030-58688-1_12

1 Introduction

1.1 Current Status of Drug Development

New drug development involves drug discovery research, non-clinical studies (animal experiments), and clinical studies, and novel drugs are manufactured and sold after approval. Animal experiments can verify the safety and therapeutic efficacy of new drugs by providing detailed information about drug metabolism and excretion. In addition, animal models of disease can be useful to elucidate underlying causes and pathological changes, providing a basis for the development of suitable drugs [1]. However, this approach is limited by the differences in drug responses between humans and animals [2, 3]. Therefore, even if the safety of a drug is confirmed by experimental animals, drug development is often stopped due to potential side effects in humans [4]. The main reasons for discontinuing drug development are side effects and toxic effects in the heart, brain, and liver [5–8]. In addition, even if the therapeutic efficacy of a drug can be verified in experimental animals, the effects may not accurately reflect those in humans and vice versa.

Owing to the above-described problems, drug development is a relatively time-consuming and costly process and is often stopped. Alternative approaches to animal experiments are highly desired for studies of drug responses and side effects.

1.2 Development of Human Induced Pluripotent Stem Cells

For drug candidate screening using culture methods, it is desirable to use primary cells derived from human tissues. However, immortalized human cells or cancer cells are generally used because cell sources are limited. Immortalized cells and cancer cells have properties different from those of tissue-derived cells, and the drug responses may therefore be different [9]. Although human embryonic stem (ES) cell-derived cells (hepatocytes and cardiomyocytes) have been used for drug screening, the cell preparation process raises ethical concerns, such as the destruction of human embryos and oocytes, because human ES cells are derived from the inner cell mass of a blastocyst in an early-stage preimplantation embryo.

In addressing these problems, human induced pluripotent stem cells (iPS cells) have been developed for use in regenerative medicine and drug discovery [10]. Human iPS cells have the capacity for differentiation into various cells, including cardiomyocytes and hepatocytes [11, 12]. Drug response tests using human iPS cells can overcome the problem related to divergence between humans and animals in non-clinical tests and can rapidly clarify the efficacy of drugs in humans. Furthermore, human disease models can be fabricated by preparing human iPS cell lines from patient cells. Human disease models are expected to contribute the development of drugs for various diseases [13].

1.3 Drug Response Using Cellular Constructs Fabricated by Tissue Engineering

Two-Dimensional (2D) Culture

The two-dimensional (2D) monolayer cell culture is a simple and rapid method to test drug responses in cells differentiated from human iPS cells. For example, the field potential duration and action potential of human iPS cell-derived cardiomyocytes can be measured using multi-electrode arrays on the bottom of a culture dish [14]. By measuring the field potential duration, it is possible to evaluate the drug response and toxicity in cardiomyocytes. In a similar manner, electrical activity due to the excitation of human iPS cell-derived neural cells can also be recorded as an extracellular local field potential by using a culture dish with electrodes [15]. Using this system, the electrophysiological characterization of drug responses in neurons can be evaluated. Additionally, human iPS cell-derived hepatocytes can be used to evaluate drug metabolism and toxicity [16].

However, the density of cells in 2D culture dishes and interactions between cells and the extracellular matrix are lower than those in organs in vivo. In addition, after human primary hepatocytes are seeded and cultured on a culture dish, albumin production and the activity of drug-metabolized enzymes are significantly reduced. Thus, characteristics and functions of a single cell differ from those of cells in 3D organs. Therefore, the fabrication of 3D tissue models is needed for pharmaceutical assays in vitro [17].

Scaffold-Based Tissue Engineering

To address the limitations of the 2D monolayer culture, extensive research has focused on the use of tissue engineering to fabricate 3D tissue models using scaffolds, such as collagen and fibrin gel. Scaffold-based tissue engineering can be divided into methods in which cells are embedded in hydrogels (such as collagen and fibrin) [18] and methods in which cells are seeded on biocompatible materials (such as polylactic acid) [19]. In contrast to the conventional 2D culture method, the scaffold-based method can provide a 3D culture environment for cells and tissue-specific functions. In a myocardial tissue model in which cardiomyocytes are embedded in a fibrin gel, it is possible to control the orientation of cardiomyocytes by electrical stimulation or mechanical stimulation and can reproduce the myocardial-specific tissue structure [20].

However, the scaffold-based 3D tissue model may interact with the drug. Nugraha et al. reported difficulties in evaluating the correct drug response of constructs because the surface of the scaffold material (such as collagen and cellulosic gel) absorbed the hydrophilic or hydrophobic drug [21]. Thus, engineered tissue models should be fabricated without a scaffold to accurately evaluate drug responses.

1.4   Bio-3D Printer Technology

A major goal is to reproduce the tissue structure using cells to accurately evaluate the drug response. Spheroid formation is a conventional method for the fabrication of a three-dimensional tissue using only cells [22]. In general, cells are seeded in low attachment plates and aggregate to form spheroids after several days. Spheroids are composed of cells and extracellular matrix produced by the cells. Importantly, cells in spheroids can be maintained at a high density three-dimensionally in vitro.

Nakayama et al. have developed bio-3D printer technology in which spheroids can be printed onto a needle array according to the desired 3D design [23]. Since the construct fabricated by a bio-3D printer does not use a scaffold, the interaction between the scaffold and drug does not occur. Several constructs, such as liver, heart, and blood vessel constructs, have been developed using a bio-3D printer [24–32].

Constructs are typically transplanted into animals to evaluate their effectiveness. The cellular constructs for blood vessels, nerve conduits, and diaphragms can be repaired by promoting tissue reconstruction in vivo [24–27]. In the case of osteo-chondral and bladder reconstructions, the constructs fabricated using mesenchymal stem cells have differentiation and regeneration ability at the transplantation site according to the surrounding environment [28, 29]. The constructs (such as the liver, heart, and trachea) are intended to function at the site of transplantation [30–32].

The constructs are useful not only for regenerative medicine but also for drug discovery. To evaluate drug responses and side effects in vitro, the cellular constructs must have tissue-specific functions. Few constructs fabricated using bio-3D printer technology have been evaluated for drug discovery. In this chapter, we introduce the fabrication method and drug response of spheroids and organoids used as ink for fabrication by a bio-3D printer. Finally, we discuss current issues and future prospects for drug discovery using cellular constructs fabricated by a bio- 3D printer.

2   Spheroid and Organoid

2.1   Spheroids and Organoids for Drug Discovery Research

To organize cells three-dimensionally without a scaffold, various techniques for spheroid and organoid fabrication have been developed. In general, both spheroids and organoids are 3D structures made of many cells. Although the terminology has been used interchangeably, there are distinct differences between these aggregate types.

Spheroids are cell aggregates formed by the promotion of cell–cell adhesion by seeding cells on a low attachment surface (Fig. 1a) [33]. Spheroids are composed of

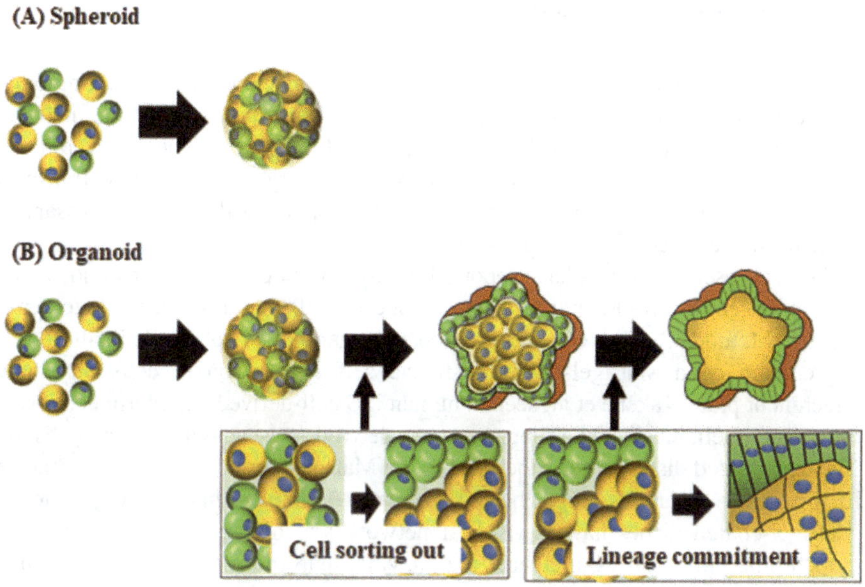

Fig. 1 Difference between spheroids and organoids. (**a**) Spheroids are formed by the self-aggregation of dissociated cells [33]. (**b**) Although organoids are initially cell aggregates, similar to spheroids, the cells in the aggregates migrate (cell sorting out) and are localized to form tissue-specific structures (lineage commitment) [34]

only cells, and cells in spheroids themselves produce the extracellular matrix in culture systems. Spheroids have tissue-specific functions.

Organoids are cell aggregates organized in vitro by stem cells, such as human iPS cells or organ progenitor cells (Fig. 1b). Organoids self-organize through cell sorting and spatially restricted lineage commitment in a manner similar to that observed in vivo [34]. Organoids are formed by the self-aggregation of cells by seeding stem cells on Matrigel consisting of a basement membrane or by embedding the cells in Matrigel. Stem cells in organoids migrate to form the tissue-specific structure [35]. Thus, organoid culture is dependent on the extracellular matrix, such as Matrigel, for formation, different from spheroids. Various organoids and spheroids, such as the liver, heart, kidney, brain, and small intestine, can be produced using human iPS cells and tissue precursor cells.

The spheroids and organoids fabricated by the above-described method are applicable for two aspects of drug development. First, they are used to search for a drug that is effective for a target disease. A disease model is prepared from human iPS cells derived from a patient cell, and the therapeutic efficacy can be evaluated by treatment with the drug candidate. The second approach is to evaluate the side effects of a drug candidate in tissues other than the target disease tissues. This section describes the fabrication method and drug responses of liver, heart, kidney, brain, and small intestine organoids or spheroids.

2.2 Hepatic Spheroids and Organoids

The liver has various functions, such as the control of blood sugar and ammonia, synthesis of various hormones, detoxification, and metabolism of drugs in vivo. Most drug molecules ($\geq$95%) absorbed in the small intestine are metabolized in the liver and are subsequently delivered to the target organ by systemic transport through blood vessels. Therefore, in order to develop a new drug, it is necessary to evaluate side effects or toxicity in the liver.

Several research groups have reported the formation of hepatic organoids using human iPS cell-derived hepatocytes or endoderm cells that differentiate into hepatocytes. Methods for the formation of hepatic organoids or spheroids include self-aggregation on a Matrigel-coated dish and self-aggregation of cells on a low attachment plate. Takebe et al. seeded human iPS cell-derived endoderm cells, vascular endothelial cells, and bone marrow-derived mesenchymal stem cells on Matrigel-coated dishes; after adherence to the Matrigel surface, they self-aggregate to form liver organoids (Fig. 2b) [36]. Hepatic organoids that self-aggregate on Matrigel-coated dishes have a vascular network and higher expression levels of cytochrome P450, a drug-metabolizing enzyme, than those in spheroids obtained from human iPS cell-derived hepatocytes alone.

Takayama et al. reported that hepatic spheroids can be formed using human iPS cells in specially processed low attachment multi-well plates (Fig. 2a) [37]. Hepatoblasts differentiated from human iPS cells are seeded on a low attachment plate and form hepatic spheroids. In general, human iPS cell-derived hepatocytes on 2D culture dishes are immature and have higher alpha-fetoprotein expression levels than those of primary hepatocytes of the adult liver. The maturation of human iPS cell-derived hepatocytes has been reported by culturing on specially processed low attachment multi-well plates [37].

The hepatotoxicity of human iPS cell-derived hepatic spheroids fabricated by the above-described method has been compared with that of spheroids of HepG2 cells (human hepatoma cells) [37]. The susceptibility of HepG2 spheroids to benzbromarone, which shows hepatotoxic effects, was lower than that of iPS cell-derived hepatic spheroids. However, the activity levels of cytochrome P450 in human iPS cell-derived hepatic spheroids were lower than those in primary hepatocytes because human iPS cell-derived hepatocytes are immature. These results indicated that human iPS cell-derived hepatic spheroids are more valuable tools for drug screening than HepG2 spheroids. Additionally, the maturation of human iPS cell-derived hepatocytes is required for use as an alternative source of primary hepatocytes in drug screening. Takebe et al. reported the maturation of hepatic organoids transplanted into mice after the administration of various stimuli, such as growth factors [38]. In the future, optimization of the maturation method for human iPS cell-derived hepatocytes is expected to provide a useful tool for evaluating drug responses.

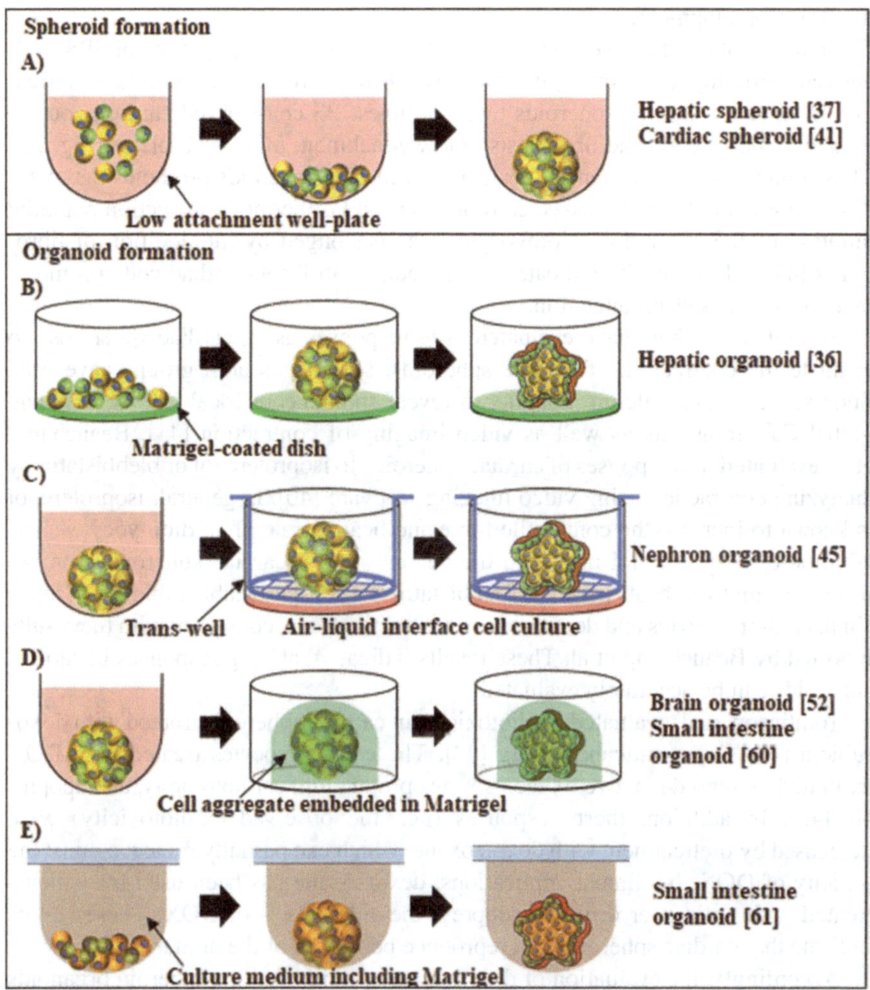

Fig. 2 Fabrication method for spheroids and organoids. (**a**) Cells are seeded in a low attachment multi-well plate. After several days, spheroids are formed by aggregating cells [37, 41]. (**b**) Cells are seeded on a Matrigel-coated dish. After adherence to the Matrigel surface, they aggregate to form an organoid [36]. (**c**) The cell aggregate is differentiated by cell culture at the air–liquid interface. After cultivation, the cell aggregate forms organoids [45, 46]. (**d**) The cell aggregate embedded in Matrigel forms organoids by stimulation with basement membrane and growth factors [52, 60]. (**e**) Using the hanging drop method, small intestine organoids can be fabricated. Matrigel is dissolved in the culture medium to form organoids [61]

2.3 Cardiac Spheroids

Heart arrhythmia is often caused by non-cardiac medications, such as antihista-mines and antipsychotics [39]. In addition, some drugs exhibit cardiotoxicity as a side effect. Thus, understanding drug response in the heart is an important aspect of new drug development.

Cardiac spheroids have been fabricated by the self-aggregation of iPS cell-derived cardiomyocytes only [40]. However, Ravenscroft et al. reported a fabrication method for cardiac spheroids that combines iPS cell-derived cardiomyocytes with endothelial cells and fibroblasts in low attachment multi-well plates (Fig. 2a). They found that the addition of fibroblasts to cardiac spheroids promotes the maturation of human iPS cell-derived cardiomyocytes. Furthermore, the action potential duration of iPS-derived cardiomyocytes was prolonged by the addition of fibroblasts [41]. These results indicate that the addition of non-cardiac cells promotes maturation and self-organization.

Several researchers have evaluated drug responses using cardiac spheroids. To evaluate the contraction of cardiac spheroids, several research groups have used fluorescence-based calcium imaging to reveal spontaneous local Ca^{2+} release and global Ca^{2+} transients as well as video imaging of contraction [41]. Beauchamp et al. evaluated the responses of cardiac spheroids to isoproterenol or blebbistatin by analyzing contraction using video imaging software [40]. In general, isoproterenol is known to increase the contractile force and beating rate of cardiomyocytes. The addition of isoproterenol increased the beating rate of cardiac spheroids from 45 beats per min to 72 beats per min. Blebbistatin is a cell-permeable inhibitor of myosin in cardiomyocytes and decreases the contractile force, consistent with the results reported by Beauchamp et al. These results indicate that drug responses in cardiac spheroids can be accurately evaluated.

Tomlinson et al. evaluated cardiotoxicity in cardiac spheroids treated with doxorubicin (DOX), an anticancer drug [42]. The cardiomyocytes treated with DOX exhibited increased reactive oxygen species production, cardiotoxicity, and apoptosis [43]. In addition, these responses (i.e., the observed cardiotoxicity) were decreased by pretreatment with dexrazoxane, which can partially protect against the toxicity of DOX. In clinical applications, dexrazoxane has been used for patients treated with anticancer drugs to suppress the side effects of DOX. These results indicate that cardiac spheroids can reproduce behaviors of the human heart.

Accordingly, the evaluation of drug responses using cardiac spheroid organoids is very useful for drug discovery owing to the ability to examine both the therapeutic efficacy and cardiotoxicity of drugs.

2.4 Nephron Organoids

The human kidney is composed of about two million nephrons and has several important functions, such as blood filtration, regulation of pH and electrolytes, and water balance in vivo. The kidney mediates the excretion of water-soluble drugs into the urine. Thus, drug toxicity in the kidney can result in renal failure and the inability to excrete waste and maintain an electrolyte balance. In phase three clinical trials, the 19% of drug development was discontinued because of nephrotoxicity [44]. Therefore, it is extremely important to understand whether a metabolized drug has toxicity or side effects in the kidney. In addition, autosomal dominant polycystic kidney disease (ADPKD) is a genetic disorder characterized by the growth of numerous cysts in both kidneys. Because the treatment for ADPKD has not been established, the development of human disease models is expected to contribute to the drug response for ADPKD.

A fabrication method for nephron organoids in vitro was developed [45]. First, human iPS cells seeded on Matrigel were cultured in medium containing CHIR99021 to induce the posterior primitive streak for 5–2 days, followed by treatment with FGF9 and heparin for another 5–2 days. After cultivation, cells were collected and spun down to form a pellet. The pellet was transferred to a Transwell membrane for culture at the air–liquid interface for 20 days (Fig. 2c). After cultivation, the pellet formed complex nephron organoids. The nephron organoid had fetal kidney-like structures and consisted of early glomeruli, proximal tubules, distal tubules, and collecting ducts [46].

Freedman et al. fabricated a nephron organoid with polycystic kidney disease from human iPS cells after genome editing [47]. Knockout of genes related to polycystic kidney disease, *PKD1* or *PKD2,* induces cyst formation from kidney tubules into nephron organoids, comparable to normal nephron organoids. The nephron organoid disease model is useful for evaluations of the therapeutic efficacy of drugs. Hale et al. evaluated the cytotoxicity of DOX using the nephron organoid [48]. The cell viability in nephron organoids decreased in a DOX concentration-dependent manner. These results indicate that toxicity in the kidney can be effectively reproduced using nephron organoids.

2.5 *Brain Organoids*

The brain can be divided into various regions, such as the forebrain, midbrain, cerebral cortex, and hippocampus. It is difficult to accurately understand the mechanisms underlying brain functions owing to complex interactions among these regions. Several research groups have studied mouse and rat brains, instead of human brains, to understand these mechanisms. In addition, the human brain is relatively large and very wrinkled compared with the mouse brain [49], making it difficult to accurately understand the mechanisms underlying human diseases involving this organ, such as schizophrenia and Alzheimer's disease, using a mouse model.

Several research groups have developed brain organoids using human iPS cells. As a general approach for the fabrication of brain organoids, Nakano et al. developed the SFEBq method [50, 51]. Human ES cells were reaggregated using a low attachment multi-well plate to form embryoid bodies (Fig. 2a). The cell aggregate was cultured in GMEM/KSR medium. After day 7, the cell aggregate was cultured in DMEM/F12/N2 for 7 days and a cerebral cortex organoid formed. Lancaster et al. generated brain organoids by embedding cells in droplets of Matrigel to provide a scaffold [52]. The embryoid bodies of human iPS cells in low attachment plates were cultured in neural induction medium to induce neuroectoderm formation. The cell aggregates were transferred to droplets of Matrigel in differentiation medium (Fig. 2d). To promote oxygen diffusion of organoids, the cell aggregates in droplets were transferred to a spinning bioreactor. In addition, the midbrain, hypothalamus, and forebrain organoid could be fabricated by changing the method and differentiation factor after differentiation into the neuroepithelium [53–55].

Several studies have evaluated neurotoxicity and drug responses in the brain organoid. The neurotoxicity of methylmercury was evaluated using brain organoids including the blood–brain barrier (BBB) structure [56]. The BBB structure prevents the entry of toxins and foreign substances from the central nervous system. A brain organoid including a BBB structure would be useful not only for neurotoxicity analyses but also for evaluating drug responses in the brain. Gonzalez et al. established human iPS cells from patients with Alzheimer's disease and formed organoids [57]. The brain organoid produced from patient-derived iPS cells had the typical abnormalities of amyloid plaques and neurofibrillary tangles. In future, these brain organoids are expected to enable studies of underlying disease mechanisms and the therapeutic efficacy of drug candidates.

2.6 Small Intestinal Organoids

The small intestine breaks down food digested in the stomach and duodenum and absorbs nutrients. The intestinal epithelium functions in the absorption of nutrients, water, and orally administered drugs. Histologically, the intestinal epithelium is composed of cilia and crypts, and the absorption and metabolism of drugs is performed by epithelial cells in the cilia [58]. Therefore, a small intestinal organoid that has cilia and crypt structures is useful for drug studies [59].

Spence et al. generated small intestinal organoids from human iPS cells [60]. Briefly, human iPS cells were treated with activin A for differentiation into the endoderm, and the treated cells were cultured in medium supplemented with Wnt3A and FGF4. The immature intestinal cells seeded on the low attachment multi-well plate-formed spheroids. These spheroids were embedded in Matrigel and cultured by the addition of EGF, Noggin, and R-spondin 1 (Fig. 2d). The small intestinal organoids consisted of columnar epithelium structures patterned into villus-like structures and crypt-like structures that expressed intestinal stem cell markers. The formation of intestinal organoids by the hanging drop culture method has recently been reported [61]. By dissolving Matrigel in the medium instead of embedding the cell aggregate in Matrigel, hanging drop culture can produce similar intestinal organoids in terms of morphology and intestinal epithelial marker expression (Fig. 2e).

Cytotoxicity in small intestinal organoids was evaluated using the anticancer drug irinotecan (CPT-11) [62]. Drug metabolism and efflux drug transporter activity in the small intestinal organoid were evaluated by adding CPT-11 adjusted to various concentrations. This study supports the use of small intestinal organoids for evaluations of toxicity.

3 Bio-3D Printing Using Spheroids or Organoids

3.1 Fabrication of Functional Tissue Constructs In Vitro

Although organoids and spheroids have been developed from human iPS cells, they cannot completely reproduce tissue-specific structures in vivo. To accurately evaluate drug responses and side effects in vitro, organoids need to have characteristics of particular tissue types, such as protein expression and structural features. Here, we describe functional tissues (liver, trachea, and heart) fabricated using a bio-3D printer.

Yanagi et al. reported that levels of albumin production and cytochrome P450 activity were higher in liver constructs fabricated by a bio-3D printer than in hepatic spheroids [31]. Kizawa et al. reported that liver constructs could maintain cytochrome P450 activity, bile duct formation, and the expression of liver-specific genes (e.g., genes involved in fatty acid synthesis, cholesterol synthesis, urea cycle, and glycolysis) at high levels [63]. Although hepatic spheroids can maintain liver-specific functions, the liver constructs can mature and maintain a high level of function by organizing spheroids with a bio-3D printer.

Taniguchi et al. developed trachea constructs using chondrocytes, endothelial cells, and mesenchymal stem cells for the purpose of tracheal reconstruction [64]. Cartilage-specific matrix components, such as aggrecan and glycosaminoglycan, were not expressed in spheroids but were observed in trachea constructs. In addition, the maturation of trachea constructs was observed after transplantation in mice.

Arai et al. and Ong et al. reported that tubular and patch-like cardiac constructs can be fabricated using human iPS cell-derived cardiomyocytes [32, 65]. Although the beat synchronization of the cardiac constructs could not be confirmed immediately after fabrication, the beat synchronization of the cardiac constructs was confirmed after cultivation. The fabricated cardiac construct also exhibited a response to electrical stimulation. Although drug responses in the cardiac construct generated using a bio-3D printer have not been reported, analyses of drug response using bio-printed structures are expected to be more accurate than that using cardiac spheroids owing to their organization.

3.2 Current Challenges and Future Prospects of Fabricating Constructs by Organoids

The practical fabrication of constructs using organoids has several limitations. This section describes current problems and future prospects for the generation of constructs by laminating organoids.

First, organoids are typically formed by embedding cell aggregates in Matrigel (Fig. 2d). Organoids embedded in Matrigel may not be used for the fabrication of constructs because they are difficult to remove. The use of low attachment plates can overcome this issue. Next, the size and shape of organoid need to be optimized for stacking using a bio-3D printer. The Kenzan method for construct fabrication involves a needle array with microneedles of 170 µm in diameter arranged at 400-µm intervals. Thus, organoids of over 500 µm in diameter are required to ensure direct contact. The optimization of organoid size is a key step in the fabrication of constructs. Finally, the total cell number used to fabricate the constructs is larger than that used to fabricate organoids, resulting in higher costs for drug discovery research. To resolve this problem, Kizawa et al. fabricated liver constructs with nine spheroids to evaluate liver-specific functions (Fig. 3a). A new type of bio-3D printer (S-PIKE) has recently been developed (see Chap. 3 for a detailed description).

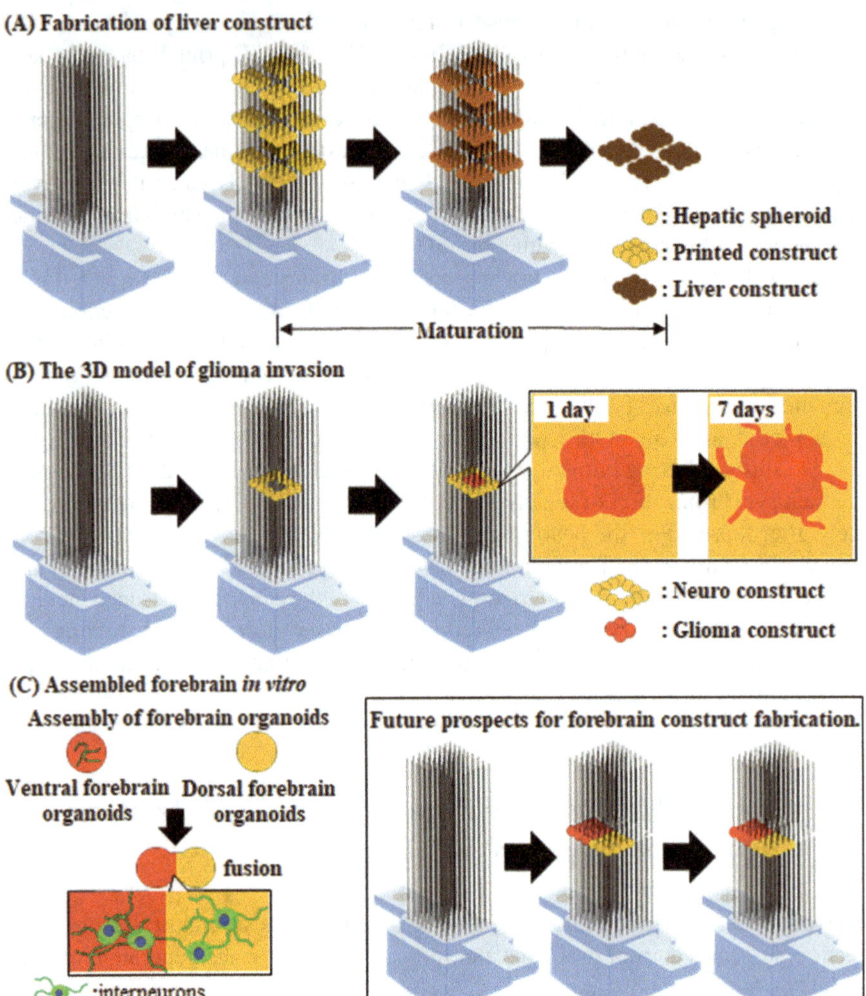

Fig. 3 Constructs for drug discovery and future prospects. (**a**) A liver construct with 9 spheroids is fabricated for drug discovery [63]. (**b**) Migration of glioma cells into nerve constructs is observed using a bio-3D printer [66]. (**c**) Left: Assembly of ventral and dorsal forebrain organoids. Interneurons in the ventral forebrain organoid migrated to dorsal forebrain organoids [67]. Right: Future prospects for forebrain tissue preparation using a bio-3D printer

Using this bio-3D printer, the arrangement of needles on the Kenzan can be freely controlled, making it possible to fabricate constructs with optimal sizes for screening drug candidates.

With respect to future prospects, the fabrication of organoid constructs has several advantages. The localization of tissue structures in constructs may be regulated by the arrangement of several kinds of organoids at the desired location using a bio-3D printer. Van Pel et al. observed the migration of glioma cells into nerve constructs by placing nerve spheroids and human glioma cell-based spheroids onto the Kenzan in a preliminary study (Fig. 3b) [66]. If the migration process of cancer cells can be reproduced, constructs will contribute substantially to the development of anticancer drugs. Birey et al. described the assembly of dorsal or ventral forebrain organoids in vitro. These subdomain-specific forebrain organoids can be assembled to recapitulate the saltatory migration of interneurons, similar to migration patterns in the fetal forebrain (Fig. 3c). Thus, the fabrication of functional constructs is expected to contribute to drug discovery in the future.

References

1. Shanks N, Greek R, Greek J (2009) Are animal models predictive for humans? Philos Ethics Humanit Med 15(4):2
2. Perel P, Roberts I, Sena E et al (2007) Comparison of treatment effects between animal experiments and clinical trials: systematic review. BMJ 334(7586):197
3. van der Worp HB, Howells DW, Sena ES et al (2010) Can animal models of disease reliably inform human studies? PLoS Med 7(3):e1000245
4. Seruga B, Ocana A, Amir E et al (2015) Failures in phase III: causes and consequences. Clin Cancer Res 21(20):4552–4560
5. Gwathmey JK, Tsaioun K, Hajjar RJ (2009) Cardionomics: a new integrative approach for screening cardiotoxicity of drug candidates. Expert Opin Drug Metab Toxicol 5:647–660
6. Ferri N, Siegl P, Corsini A et al (2013) Drug attrition during pre-clinical and clinical development: understanding and managing drug-induced cardiotoxicity. Pharmacol Ther 138:470–484
7. Katarey D, Verma S (2016) Drug-induced liver injury. Clin Med (Lond) 16(Suppl 6):s104–s109
8. Gao Z, Chen Y, Cai X et al (2017) Predict drug permeability to blood-brain-barrier from clinical phenotypes: drug side effects and drug indications. Bioinformatics 33(6):901–908
9. Astashkina AI, Mann BK, Prestwich GD et al (2012) Comparing predictive drug nephrotoxicity biomarkers in kidney 3-D primary organoid culture and immortalized cell lines. Biomaterials 33(18):4712–4721
10. Takahashi K, Tanabe K, Ohnuki M et al (2007) Induction of pluripotent stem cells from adult human fibroblasts by defined factors. Cell 131(5):861–872
11. Si-Tayeb K, Noto FK, Nagaoka M et al (2010) Highly efficient generation of human hepatocyte-like cells from induced pluripotent stem cells. Hepatology 51(1):297–305
12. Zwi L, Caspi O, Arbel G et al (2009) Cardiomyocyte differentiation of human induced pluripotent stem cells. Circulation 120(15):1513–1523
13. Mungenast AE, Siegert S, Tsai LH (2016) Modeling Alzheimer's disease with human induced pluripotent stem (iPS) cells. Mol Cell Neurosci 73:13–31
14. Asakura K, Hayashi S, Ojima A et al (2015) Improvement of acquisition and analysis methods in multi-electrode array experiments with iPS cell-derived cardiomyocytes. J Pharmacol Toxicol Methods 75:17–26

15. Odawara A, Matsuda N, Ishibashi Y et al (2018) Toxicological evaluation of convulsant and anticonvulsant drugs in human induced pluripotent stem cell-derived cortical neuronal networks using an MEA system. Sci Rep 8(1):10416

16. Kang SJ, Lee HM, Park YI et al (2016) Chemically induced hepatotoxicity in human stem cell-induced hepatocytes compared with primary hepatocytes and HepG2. Cell Biol Toxicol 32(5):403–417

17. Pampaloni F, Reynaud EG, Stelzer EH (2007) The third dimension bridges the gap between cell culture and live tissue. Nat Rev Mol Cell Biol 8(10):839–845

18. Hansen A, Eder A, Bönstrup M et al (2010) Development of a drug screening platform based on engineered heart tissue. Circ Res 107(1):35–44

19. Kasuya J, Sudo R, Tamogami R et al (2012) Reconstruction of 3D stacked hepatocyte tissues using degradable, microporous poly(d,l-lactide-co-glycolide) membranes. Biomaterials 33(9):2693–2700

20. Stoppel WL, Kaplan DL, Black LD (2016) Electrical and mechanical stimulation of cardiac cells and tissue constructs. Adv Drug Deliv Rev 96:135–155

21. Nugraha B, Hong X, Mo X et al (2011) Galactosylated cellulosic sponge for multi-well drug safety testing. Biomaterials 32(29):6982–6994

22. Lin RZ, Chang HY (2008) Recent advances in three-dimensional multicellular spheroid culture for biomedical research. Biotechnol J 3(9–10):1172–1184

23. Moldovan NI, Hibino N, Nakayama K (2017) Principles of the Kenzan method for robotic cell spheroid-based three-dimensional bioprinting. Tissue Eng Part B Rev 23(3):237–244

24. Itoh M, Mukae Y, Kitsuka T et al (2019) Development of an immunodeficient pig model allowing long-term accommodation of artificial human vascular tubes. Nat Commun 10(1):2244

25. Yurie H, Ikeguchi R, Aoyama T et al (2017) The efficacy of a scaffold-free Bio 3D conduit developed from human fibroblasts on peripheral nerve regeneration in a rat sciatic nerve model. PLoS One 12(2):e0171448

26. Mitsuzawa S, Ikeguchi R, Aoyama T et al (2019) The efficacy of a scaffold-free bio 3D conduit developed from autologous dermal fibroblasts on peripheral nerve regeneration in a canine ulnar nerve injury model: a preclinical proof-of-concept study. Cell Transplant 28(9–10):1231–1241

27. Zhang XY, Yanagi Y, Sheng Z et al (2018) Regeneration of diaphragm with bio-3D cellular patch. Biomaterials 167:1–14

28. Imamura T, Shimamura M, Ogawa T et al (2018) Biofabricated structures reconstruct functional urinary bladders in radiation-injured rat bladders. Tissue Eng Part A 24(21–22):1574–1587

29. Yamasaki A, Kunitomi Y, Murata D et al (2019) Osteochondral regeneration using constructs of mesenchymal stem cells made by bio three-dimensional printing in mini-pigs. J Orthop Res 37(6):1398–1408

30. Itoh M, Nakayama K, Noguchi R et al (2015) Scaffold-free tubular tissues created by a Bio-3D printer undergo remodeling and endothelialization when implanted in rat aortae. PLoS One 10(9):e0136681

31. Yanagi Y, Nakayama K, Taguchi T et al (2017) In vivo and ex vivo methods of growing a liver bud through tissue connection. Sci Rep 7(1):14085

32. Ong CS, Fukunishi T, Zhang H et al (2017) Biomaterial-free three-dimensional bioprinting of cardiac tissue using human induced pluripotent stem cell derived cardiomyocytes. Sci Rep 7(1):4566

33. Sutherland RM, McCredie JA, Inch WR (1971) Growth of multicell spheroids in tissue culture as a model of nodular carcinomas. J Natl Cancer Inst 46(1):113–120

34. Lancaster MA, Knoblich JA (2014) Organogenesis in a dish: modeling development and disease using organoid technologies. Science 345(6194):1247125

35. Takebe T, Wells JM (2019) Organoids by design. Science 364(6444):956–959

36. Takebe T, Enomura M, Yoshizawa E et al (2015) Vascularized and complex organ buds from diverse tissues via mesenchymal cell-driven condensation. Cell Stem Cell 16(5):556–565

37. Takayama K, Kawabata K, Nagamoto Y et al (2013) 3D spheroid culture of hESC/hiPSC-derived hepatocyte-like cells for drug toxicity testing. Biomaterials 34(7):1781–1789

38. Takebe T, Sekine K, Enomura M et al (2013) Vascularized and functional human liver from an iPSC-derived organ bud transplant. Nature 499(7459):481–484

39. Heist EK, Ruskin JN (2010) Drug-induced arrhythmia. Circulation 122(14):1426–1435

40. Beauchamp P, Moritz W, Kelm JM et al (2014) Development and characterization of a scaffold-free 3D spheroid model of iPSC-derived human cardiomyocytes. Tissue Eng Part C Methods 21(8):852–861

41. Ravenscroft SM, Pointon A, Williams AW et al (2016) Cardiac non-myocyte cells show enhanced pharmacological function suggestive of contractile maturity in stem cell derived cardiomyocyte microtissues. Toxicol Sci 152(1):99–112

42. Tomlinson L, Lu ZQ, Bentley RA et al (2019) Attenuation of doxorubicin-induced cardio-toxicity in a human in vitro cardiac model by the induction of the NRF-2 pathway. Biomed Pharmacother 112:108637

43. Shimauchi T, Numaga-Tomita T, Ito T et al (2017) TRPC3-Nox2 complex mediates doxorubicin-induced myocardial atrophy. JCI Insight 2(15). pii: 93358

44. Ran S, Li Y, Zink D et al (2014) Supervised prediction of drug-induced nephrotoxicity based on interleukin-6 and -8 expression levels. BMC Bioinformatics 15(Suppl 16):S16

45. Takasato M, Er PX, Becroft M et al (2014) Directing human embryonic stem cell differentiation towards a renal lineage generates a self-organizing kidney. Nat Cell Biol 16(1):118–126

46. Takasato M, Er PX, Chiu HS et al (2016) Kidney organoids from human iPS cells contain multiple lineages and model human nephrogenesis. Nature 536(7615):238

47. Freedman BS, Brooks CR, Lam AQ et al (2015) Modelling kidney disease with CRISPR-mutant kidney organoids derived from human pluripotent epiblast spheroids. Nat Commun 6:8715

48. Hale LJ, Howden SE, Phipson B et al (2018) 3D organoid-derived human glomeruli for personalised podocyte disease modelling and drug screening. Nat Commun 9(1):5167

49. Sun T, Hevner RF (2014) Growth and folding of the mammalian cerebral cortex: from molecules to malformations. Nat Rev Neurosci 15(4):217–232

50. Nakano T, Ando S, Takata N et al (2012) Self-formation of optic cups and storable stratified neural retina from human ESCs. Cell Stem Cell 10(6):771–785

51. Eiraku M, Watanabe K, Matsuo-Takasaki M (2008) Self-organized formation of polarized cortical tissues from ESCs and its active manipulation by extrinsic signals. Cell Stem Cell 3(5):519–532

52. Lancaster MA, Renner M, Martin CA et al (2013) Cerebral organoids model human brain development and microcephaly. Nature 501(7467):373–379

53. Sakaguchi H, Ozaki Y, Ashida T et al (2019) Self-organized synchronous calcium transients in a cultured human neural network derived from cerebral organoids. Stem Cell Rep 13(3):458–473

54. Ogura T, Sakaguchi H, Miyamoto S et al (2018) Three-dimensional induction of dorsal, intermediate and ventral spinal cord tissues from human pluripotent stem cells. Development 145(16). pii: dev162214

55. Qian X, Jacob F, Song MM et al (2018) Generation of human brain region-specific organoids using a miniaturized spinning bioreactor. Nat Protoc 13(3):565–580

56. Nzou G, Wicks RT, Wicks EE et al (2018) Human cortex spheroid with a functional blood brain barrier for high-throughput neurotoxicity screening and disease modeling. Sci Rep 8(1):7413

57. Gonzalez C, Armijo E, Bravo-Alegria J et al (2018) Modeling amyloid beta and tau pathology in human cerebral organoids. Mol Psychiatry 23(12):2363–2374

58. Billat PA, Roger E, Faure S et al (2017) Models for drug absorption from the small intestine: where are we and where are we going? Drug Discov Today 22(5):761–775

59. Fair KL, Colquhoun J, Hannan NRF (2018) Intestinal organoids for modelling intestinal development and disease. Philos Trans R Soc Lond Ser B Biol Sci 373(1750). pii: 20170217

60. Spence JR, Mayhew CN, Rankin SA et al (2011) Directed differentiation of human pluripotent stem cells into intestinal tissue in vitro. Nature 470(7332):105–109
61. Panek M, Grabacka M, Pierzchalska M (2018) The formation of intestinal organoids in a hanging drop culture. Cytotechnology 70(3):1085–1095
62. Lu W, Rettenmeier E, Paszek M et al (2017) Crypt organoid culture as an in vitro model in drug metabolism and cytotoxicity studies. Drug Metab Dispos 45(7):748–754
63. Kizawa H, Nagao E, Shimamura M et al (2017) Scaffold-free 3D bio-printed human liver tissue stably maintains metabolic functions useful for drug discovery. Biochem Biophys Rep 10:186–191
64. Taniguchi D, Matsumoto K, Tsuchiya T et al (2018) Scaffold-free trachea regeneration by tissue engineering with bio-3D printing. Interact Cardiovasc Thorac Surg 26(5):745–752
65. Arai K, Murata D, Verissimo AR et al (2018) Fabrication of scaffold-free tubular cardiac constructs using a Bio-3D printer. PLoS One 13(12):e0209162
66. van Pel DM, Harada K, Song D et al (2018) Modelling glioma invasion using 3D bioprinting and scaffold-free 3D culture. J Cell Commun Signal 12(4):723–730
67. Birey F, Andersen J, Makinson CD et al (2017) Assembly of functionally integrated human forebrain spheroids. Nature 545(7652):54–59

Biomaterial-Related Surgical Site Infection: Anti-infectious Metal Coating on Biomaterials

Hironobu Koseki and Shiro Kajiyama

Abstract Artificial biomaterials show disadvantages in terms of local bacterial infection and carry a risk of extremely refractory implant-related surgical site infections (SSIs). Pathogenic bacteria, which are commonly staphylococci, form a biofilm on the surface of biomaterial and acquire increased resistance to biological immune mechanisms and antibiotics. The economic burden of treatments for only postoperative prosthetic hip or knee infection (prosthetic joint infection, PJI) in the United States is anticipated to reach \$1.6 billion by 2020. Inorganic antibacterial materials using silver, antibiotic agents, and iodine (disinfectant) as well as titanium dioxide (TiO_2) are the focus of research into countermeasures against implant-related SSI. The antibacterial activity of TiO_2 is not species-specific, does not induce drug-resistant strains, and does not prevent osteogenesis. An understanding of the risks of bacterial infection with biomaterials is crucial in regenerative medicine. The Kenzan method is expected to be superior for preventing implant-related infection in regenerative medicine.

Keywords Biomaterial · Implant-related infection · Titanium dioxide

H. Koseki (✉)
Department of Health Sciences, Nagasaki University Graduate School of Biomedical Sciences, Nagasaki, Japan
e-mail: koseki@nagasaki-u.ac.jp

S. Kajiyama
Department of Orthopedic Surgery, Nagasaki University Graduate School of Biomedical Sciences, Nagasaki, Japan

© Springer Nature Switzerland AG 2021
K. Nakayama (ed.), *Kenzan Method for Scaffold-Free Biofabrication*,
https://doi.org/10.1007/978-3-030-58688-1_13

1   Introduction

Innovations in surgical techniques and the development and improvement of biomaterials for medical use have greatly contributed to advances in medicine to date. In current medical practice, surgical biomaterials (bio-implants) play crucial roles by virtue of their low toxicity, low corrosion rates, and favorable mechanical properties. As a result, various products are available for a wide variety of purposes. In the field of orthopedic surgery, devices composed of titanium, stainless steel, cobalt-chromium alloy, or ultrahigh molecular weight polyethylene are frequently implanted into human bones and joints. These tissues, however, are essentially sterile in the native state and contain very few phagocytes, and host defense mechanisms are therefore naturally hard to function. Moreover, since angiogenesis does not occur in these artificial materials themselves, they are at an extreme disadvantage in terms of local bacterial infections and carry a risk of surgical site infection (SSI). Bacteria adhering to and proliferating on the surface of biomaterials produce a matrix of extracellular polysaccharides (EPSs) such as glycoproteins over time to surround the bacterial cells. This structure wrapped by polysaccharide-containing matrix produced by bacteria is called biofilm [1]. Since bacteria protected by biofilm acquire increased resistance to biological immune mechanisms and antibiotics [1–3], treatment of implant-related SSI becomes much more difficult and typically involves pharmacotherapy with antibiotics at doses as high as the physical status of the patient allows. For deep infections, surgical treatments such as washing and debridement become necessary. Surgeons often exert major efforts in attempting to control infections, unavoidably involving removal of the implant and continuous washing or placement of antibiotic-containing bio-cement. Moreover, even if multiple organ failure or death from sepsis can be circumvented, the physical disabilities arising from serial treatments of infection can be considerable, since the SSI ultimately require multiple surgeries, large medical expenses, a long period of hospitalization, and significant pain. To date, careful preventive measures against SSI have included methods for disinfecting the surgical wound, sterilization of devices and instruments, and use of bioclean rooms. However, SSI still occur in 0.2–17.3% of all orthopedic surgeries today [4, 5], and new preventive methods applying multifaceted approaches are highly desired.

2   History of Implant-Related Infections

In the early 1970s, the Centers for Disease Control and Prevention (CDC) in the United States created the National Nosocomial Infections Surveillance (NNIS) system. This system sought to clarify the status of hospital infections such as risk factors and causative bacteria and reported 15,523 SSIs following 593,344 operations between 1986 and 1996, with 77% of deaths following complications from surgery reportedly related to SSI [6]. Furthermore, the incidence of postoperative prosthetic hip or knee infection (prosthetic joint infection, PJI) in the United States is reported to be 2.0–2.4% and is increasing every year, while the economic burden of treatments for these infections is expected to reach \$1.6 billion by 2020 [7]. Under such circumstances, the CDC presented the Guideline for Prevention of Surgical Site Infection in 1999 [6], and the first edition of post-bone/joint surgery infection prevention guidelines in Japan was published in 2006. Moreover, the Infectious Diseases Society of America (IDSA) announced clinical practice guidelines for the diagnosis and treatment of PJI in 2013 [8], while the CDC has been globally involved in the reduction of implant-related infection incidences through multiple revisions of the Guideline for the Prevention of Surgical Site Infection [9]. However, based on the facts that (1) SSI occurs in 1.5–2.5% of all primary joint arthroplasties, and this percentage exceeds 20% for revision procedures [10]; and (2) 572,000 total hip arthroplasties and 3,480,000 knee arthroplasties are projected to be performed annually in the United States by 2030, we can estimate the occurrence of over approximately 80,000 cases of SSI annually [11, 12]. Furthermore, in association with longer life spans and an increasingly elderly population, cases at high risk of SSI due to decreased immune function, comorbidities such as diabetes mellitus, and immunodeficiency due to immunosuppressants such as steroids represent a matter of grave concern.

3 Pathogenic Bacteria of SSI and Route of Infection

With the aging of society and advances in medical care, low-virulence bacterial infections that occur in immunosuppressed hosts have been a significant issue for many years. The most common causative microorganisms for SSI following prosthetic arthroplasty are considered to be methicillin-sensitive *Staphylococcus aureus* (MSSA), methicillin-resistant *S. aureus* (MRSA), and coagulase-negative staphylococci [13–15]. The main concern in recent years has been the increase in infections due to antibiotic-resistant bacteria. Implant-related infection with MRSA poses a particularly serious threat, as these strains are particularly resistant to antibiotics. Salgado et al. observed that PJI due to MRSA carries a higher risk of treatment failure than PJI caused by MSSA [16]. Routes of infection for SSI are largely divisible into the "internal path" from a different site of the body such as the nasal cavity and the "external path" through wound contamination from outside the body. In particular, intraoperative airborne bacteria and bacterial contamination of the surgical wound have been considered as the leading causes of implant-related infections [15, 17, 18].

4 SSI Case Presentations

Case 1: A 69-year-old man with diabetes mellitus underwent hip hemiarthroplasty for treatment of a left femoral neck fracture, but systemic fever and pain around the left hip worsened 6 months later. Inflammatory markers were positive (C-reactive protein (CRP), 3.4 mg/dL; erythrocyte sedimentation rate (ESR), 69 mm/h), and plain X-ray images showed osteolytic signs around the sunken femoral stem (Fig. 1). Under a diagnosis of PJI in the left hip, immediate surgical treatments including

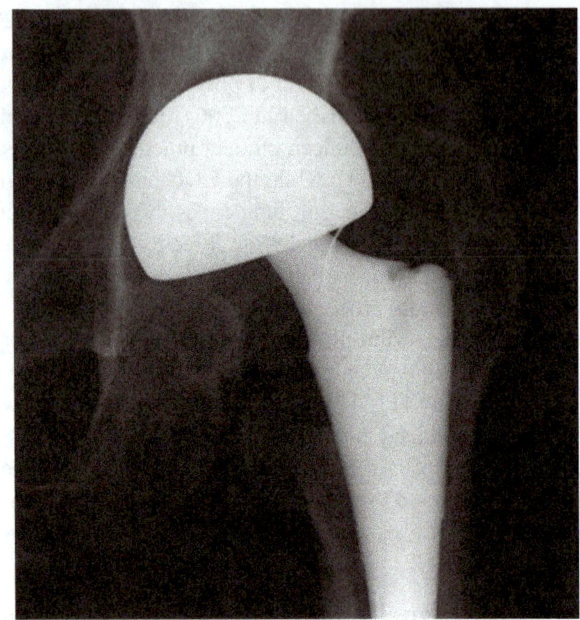

Fig. 1 Case 1; Plain X-ray at initial examination **Osteolysis around the stem and shortening of leg length due to a sunken stem are observed**

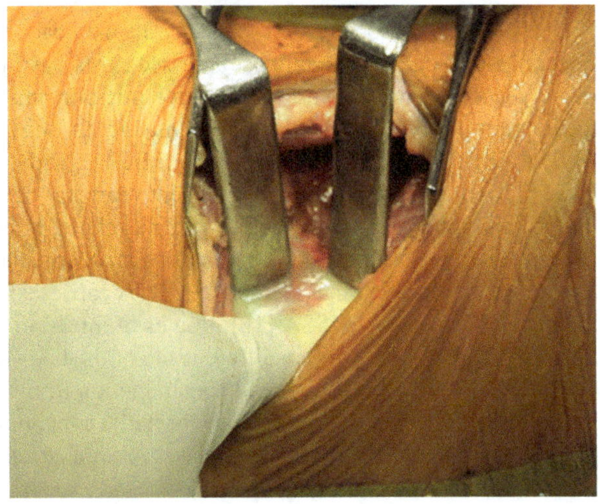

Fig. 2 Case 1; Intraoperative finding **When the wound is opened, discharge of white, viscous pus from the deep area is observed**

removal of the prosthetic head of the femoral stem, debridement, and placement of antibiotic-containing cement molds and beads were performed. In the joint, white purulent fluid and hyperplasia of reddish-brown, thickened synovial tissue were evident (Fig. 2).

MRSA was detected in cultured samples of fluid and synovial tissue, and anti-MRSA therapy was started. Two months later, the second surgery for PJI was performed, but infection had not yet subsided. Debridement was therefore performed, and antibiotic-containing cement molds and beads were again embedded (Fig. 3). Four months after the initial surgery, as infection had subsided according to clinical

Fig. 3 Case 1; Plain X-ray after second debridement operation
The stem is not loose and cannot be removed. After adequate debridement and synovectomy, antibiotic-containing cement molds and beads are placed

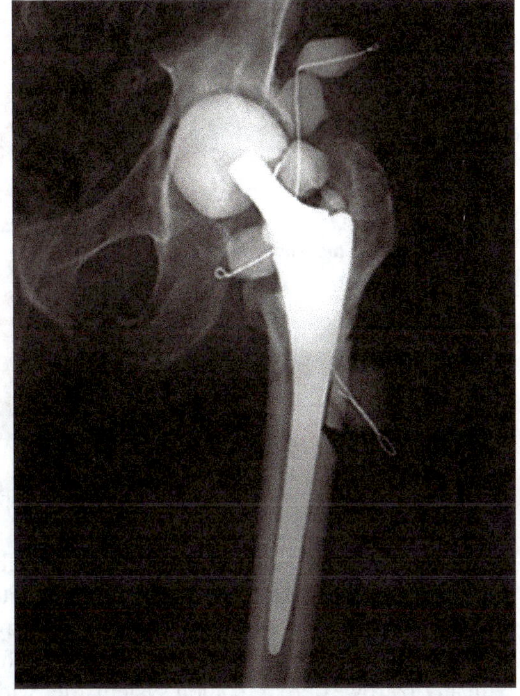

Fig. 4 Case 2; Macroscopic findings at transport to the emergency room
Open comminuted fracture of the right lower thigh is evident, and a bone fragment is exposed through the wound (Gustilo IIIb)

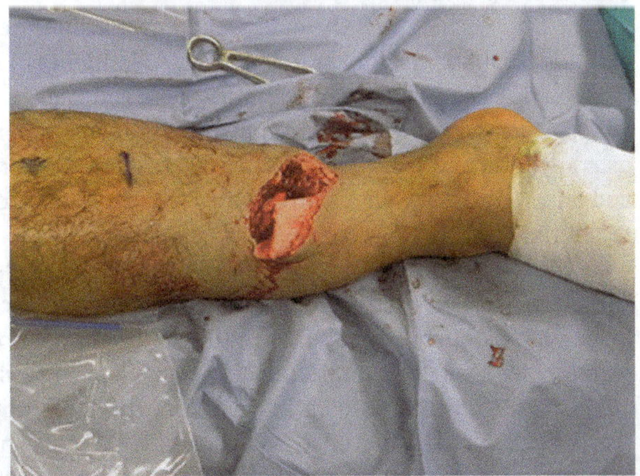

findings and intraoperative rapid pathology examination, a new prosthetic joint was implanted after careful curettage of bone and soft tissue. Fortunately, no infection has recurred since. However, due to long-term recumbency, muscle atrophy and joint contracture have resulted in the patient requiring a cane to walk.

Case 2: A 43-year-old man was transported via ambulance due to open comminuted fracture of the right lower thigh (Gustilo IIIb) sustained in a traffic accident (Fig. 4). He immediately underwent washing, debridement, and external fixation, and then careful antibiotic pharmacotherapy was provided. Eight days later, tibial

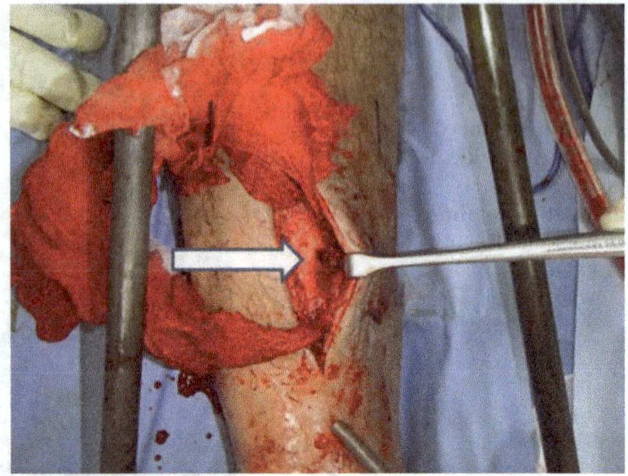

Fig. 5 Case 2; Pin site infection at 3 months after injury
Exudate discharge and frank purulence are observed from the insertion area of the distal pin for external fixation, and fistula through the bone marrow is observed (arrow)

intramedullary nail fixation and vascularized flap were performed. However, frank purulence from the insertion area of the distal pin for external fixation (pin site infection) was observed 3 months later. Although a broad-spectrum antibiotic was administered via drip infusion, MRSA was detected from pus 4 months after injury. The patient was diagnosed with pyogenic osteomyelitis caused by pin site infection. The intramedullary nail was removed, and the fractured bones were instead fixed with an Ilizarov external fixation system after wide-ranging debridement, including bone tissue (Fig. 5).

The patient subsequently underwent repeated bone debridement and exchange of antibiotic-containing cement beads every 2–3 months. Septic shock was able to be prevented, but ultimately 18 months of period with a total of nine surgeries (of which seven were procedures against osteomyelitis) were required for treatment of osteomyelitis and bone union. Due to restricted range of motion of the right knee and ankle joints as well as atrophy of the right leg muscles, the patient was unable to return to his previous work in physical labor.

Case 3: A 75-year-old woman, who had developed rheumatoid arthritis 20 years earlier, underwent right knee prosthetic arthroplasty for rheumatoid arthritis of the knee. On postoperative day 9, swelling and redness around the surgical wound as well as a fever of 40 °C appeared along with a positive inflammatory marker (CRP: 17.43 mg/dL). Since purulent, turbid joint fluid was obtained from arthrocentesis, the patient immediately underwent operation against pyogenic arthritis including extensive curettage and exchange of the insert component. In the joint, white purulent synovial fluid was present, and hyperplasia of reddish-brown, thickened synovial tissue was evident (Fig. 6). MRSA was detected in culture samples of joint fluid and synovial tissue, and anti-MRSA agents were immediately given via drip infusion. Subsequently, hypoproteinemia (albumin, 2.4 g/dl) and anemia (hemoglobin, 6.7 g/dl) developed, which required administration of albumin formulation and blood transfusion. In addition, acute renal failure developed as a side effect of pharmacotherapy, and the drugs were inevitably continually switched. Surgical debridement was performed every other week for a total of three times. After 3 months,

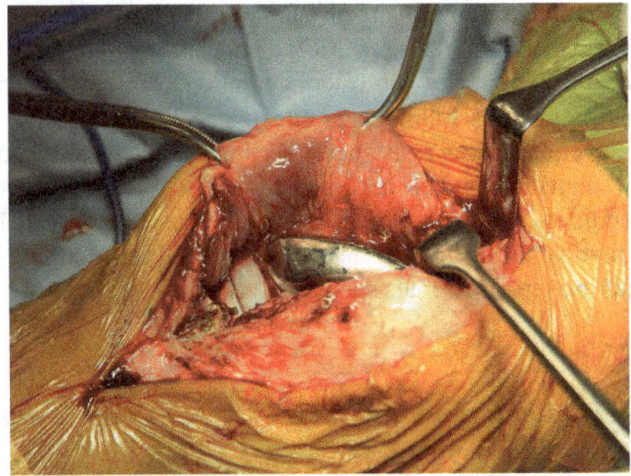

Fig. 6 Case 3; Macroscopic findings at first debridement surgery **With the opening of the wound, white purulent synovial fluid is released, and hyperplasia of reddish-brown, thickened synovial tissue is evident. Wide-ranging curettage and exchange of components are performed**

vascularized flap surgery was performed for the skin defect. No infection has recurred through continuation of antibiotics for 6 months, but the patient remains unable to walk and requires a wheelchair for mobility.

5 Inorganic Antibacterial Materials as Innovative Preventive Measures Against SSI

Medical biomaterials have made immeasurable contributions to medicine. With future advances in biomaterials engineering, various bio-implants will continue to be developed and play ever more critical roles. However, implant-related infections, as serious and devastating complications, need to be resolved. Recently, inorganic antibacterial materials using silver, antibiotic agents, and iodine (disinfectant) have been gaining major attention. However, as they continue to release such bioactive agents in the human body, accumulation in organs has been a concern in terms of potential negative effects in living tissues.

We have studied the application of photocatalytic titanium dioxide (TiO_2) as a coating on the surface of biomaterials in an attempt to generate antibacterial and bactericidal properties in the implant biomaterials themselves that are disadvantageous for infections. The photocatalyst TiO_2 has three primary crystal types – rutile, anatase, and brookite – and various properties can be expressed depending on the method and conditions of creation. When anatase-type crystals are exposed to ultraviolet (UV) light, active oxygen free radicals such as $\cdot OH$, O_2^-, HO_2^-, and H_2O_2 are released. The oxidation potential produced with these substances is stronger than those of chlorine or hypochlorous acid and is capable of degrading organic substances such as bacteria and viruses. TiO_2 is chemically stable and nontoxic and so has been used in environmental fields such as the outer walls of buildings, antibacterial masks, and water/air purification [19]. The surface of pure titanium implants currently used in orthopedic surgeries becomes oxidized naturally when exposed to air, resulting in the same TiO_2 composition, which is known to be bio-inert and to enable bioactivity for osteointegration. The characteristics of TiO_2 bactericidal activity are as follows: (1) effectiveness regardless of the bacterial strain, (2) no induction of drug resistance, (3) adjustment by manipulating the type and amount of illumination, and (4) decomposition of toxic debris as well as bacteria itself [20]. In addition, TiO_2 does not prevent normal osteogenesis [21] and does not continue to dissolve or accumulate in the human body, which are superior to other inorganic antibacterial materials such as silver, antibiotic agents, and iodine.

5.1 Bactericidal Effects of a Thin TiO_2 Film Against S. aureus [22]

TiO_2 film was created using the plasma source ion implantation (PSII) method, which is capable of implanting ions and depositing films uniformly and completely over the surfaces of three-dimensional targets at a reasonable cost. Scanning electron micrograph (SEM) images of the surface of TiO_2 film show innumerable globular fine microstructures with a consistent particle size of 0.1–0.5 µm homogeneously throughout the sample. This finding was absent on the surface of basic materials (Fig. 7a, b).

Commercially pure titanium (ASTM F-67) and stainless steel (SUS316) substrates for medical use were coated with a thin TiO_2 film (film thickness, <1 µm). S. aureus (strain Seattle 1945: ATCC 25923) was incubated and suspended at a concentration of 1×10^5 cells/mL, and then the bacterial suspension was dispensed onto each substrate (solution pH 7.0). UV-A (intensity, 2.0 mW/cm^2; peak wavelength, 352 nm) was illuminated with a black light, and viability was calculated by

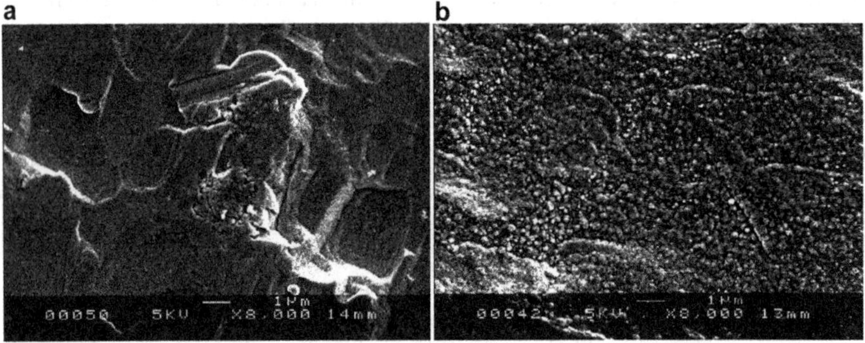

Fig. 7 SEM micrographs of the surfaces of (**a**) untreated and (**b**) TiO_2-coated samples
TiO_2 photocatalytic nanoparticles adhere to the surface of TiO_2-coated substrate. Scale bar = 1 µm. Original magnification ×8000

measuring the viable bacterial count using the plate dilution method. Bacterial viability was calculated as follows:

$$\text{Viability at } \alpha \text{ minutes} (\%)$$
$$= \text{colony forming units} (\text{CFU}) \text{ at } \alpha \text{ minutes} / \text{CFU at } 0 \text{ minutes} \times 100$$

In the UV-A-only group (Group 3), the viability of *S. aureus* declined gradually as UV-A illumination proceeded due to the bactericidal properties of the UV rays themselves. However, the finding of interest here was the suppression of *S. aureus* viability for TiO_2 + UVA samples (Group 1) from the early phase of UVA illumination in both types of biomaterial. Significant differences in viability were seen between Group 1 and the other groups at 30 and 45 min of UV illumination ($P < 0.05$) (Fig. 8a, b).

These results indicate that the photocatalytic bactericidal action of the TiO_2 film was effectively expressed against *S. aureus*, suggesting the potential to reduce the incidence of SSI associated with biomaterials. Even if bacteria adhere to the TiO_2 deposited prosthetic biomaterial surface during the joint arthroplasty, adherent bacteria can be killed by UV-A illumination and may serve to restore surfaces to a bacterium- and toxin-free state. Furthermore, TiO_2 characteristically provides no photocatalytic action unless exposed to light. This means that any TiO_2 within the body after closing the wound would become chemically inactive and would have no negative effects on the physiological environment. These points are clearly different from metal surfaces treated with silver ion products [23] or antibiotic agents [24].

5.2 Effects on Preventing Pin Site Infection for External Fixation [25]

The efficacy of stainless steel (SUS316L) screw pins coated with TiO_2 for the inhibition of infection was compared with that of untreated pins in an in vivo study. TiO_2-coated and untreated screw pins were inserted into the femoral bones of 7-week-old Sprague-Dawley (SD) specific pathogen-free (SPF) rats, followed by bacterial contamination with *S. aureus* at 1×10^8 cells/mL. After 30 min of illumination with UV-A (intensity, 2.0 mW/cm^2; peak wavelength, 352 nm), macroscopic clinical findings [26] and bone histomorphology [24] were assessed at 2 weeks. Clinical signs of infection were present in 23 of 30 untreated pins (76.7%), compared to only 11 of 30 TiO_2-coated pins (36.7%) (Fig. 9a–c, Table 1).

The histological bone infection score and planimetric rate of occupation for bacterial colonies and neutrophils in the TiO_2 pin group were significantly lower than those in the untreated pin group. Bone-implant contact ratio was significantly greater in the TiO_2 pin group (71.4%) than in the untreated pin group (58.2%; $P < 0.05$). Furthermore, the area surrounding the TiO_2-coated pins showed almost none of the findings observed for the untreated pins, such as severe bone resorption

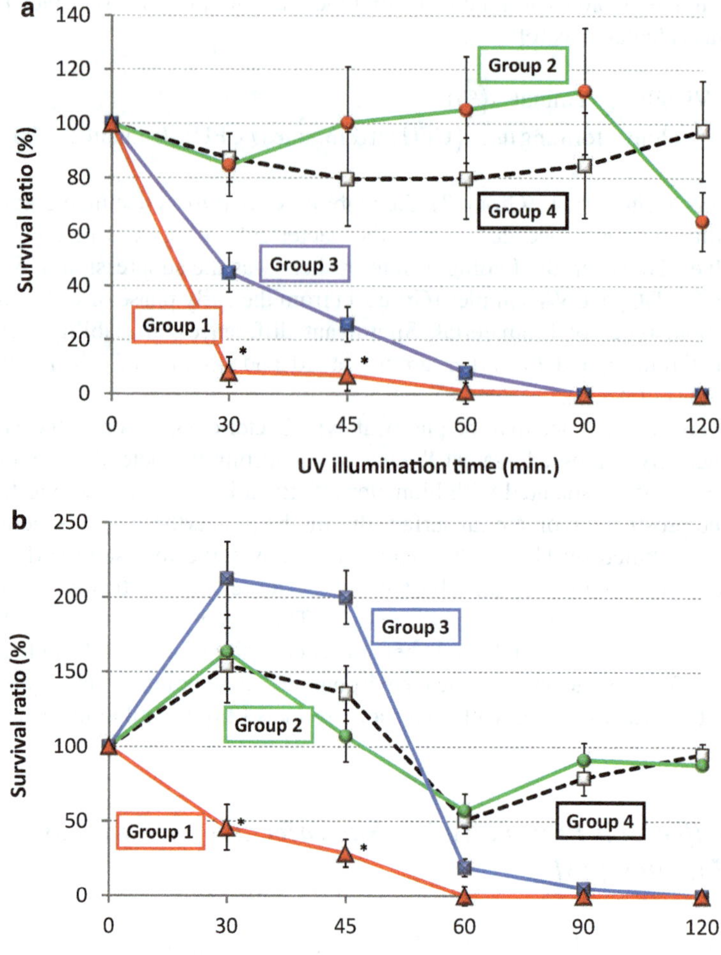

Fig. 8 Photocatalytic inactivation of *S. aureus* on (**a**) pure titanium and (**b**) stainless steel
With 30 min and 45 min of UV illumination, bacterial viability is significantly suppressed in Group 1 compared to other groups
* ANOVA, $P < 0.05$; significantly different from the other three groups, Group 1, with TiO_2 film + UV-A illumination; Group 2, with TiO_2 film + UV-A shield; Group 3, without TiO_2 film + UV-A illumination; Group 4, without TiO_2 film + UV-A shield
Complete bacterial inactivation on the pure titanium was achieved after 90 min, versus 60 min on stainless steel

and destruction, abscesses due to bacteria, and inflammatory cell migration caused by bacterial infection (Table 2). Moreover, the TiO_2-coated pins were in direct contact with bone tissue without formation of a pseudo-film, and both angiogenesis and osteogenesis were occurring, suggesting high biocompatibility of TiO_2 (Fig. 10a, b).

Pin site infection is the most common and significant complication of external fixation. Reported rates of pin site infection vary widely in the literature, ranging

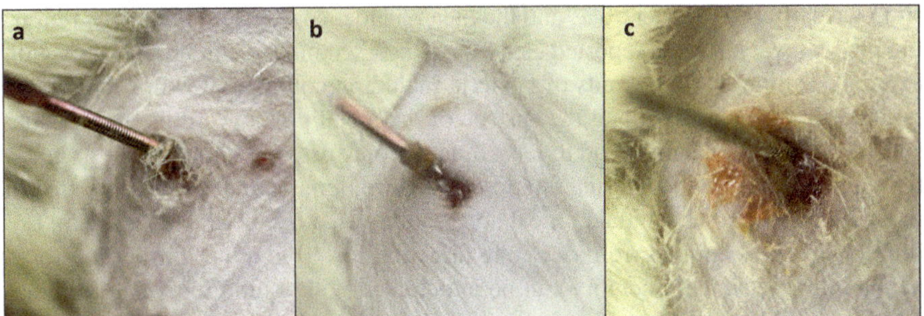

Fig. 9 Classification of macroscopic findings
(**a**) Without signs of infection, (**b**) with signs of inflammation/drainage, (**c**) with pus discharge

Table 1 Evaluation of clinical findings according to criteria for pin site infection

	Infection (−)	Inflammation, drainage	Pus discharge	Total
Untreated pin	7 (23.3%)	16 (53.3%)	7 (23.4%)	30
TiO_2 pin	19 (63.3%)	9 (30.0%)	2 (6.7%)	30

Signs of infection, purulence, and/or serous drainage were significantly more common in the control pin group (76.7%) than in the TiO_2-coated pin group (36.7%, P < 0.01)

Table 2 Histomorphometric analysis presented as mean score ± standard deviation

	Bone infection score (points)	Bone implant contact ratio (%)	Area of colonies and neutrophils (%)
Untreated pin	4.9 ± 1.0	58.2 ± 8.1	24.7 ± 10.3
TiO_2 pin	3.1 ± 1.6[a]	71.4 ± 5.4[a]	13.3 ± 6.4[a]

[a]Mann-Whitney U-test; P < 0.01

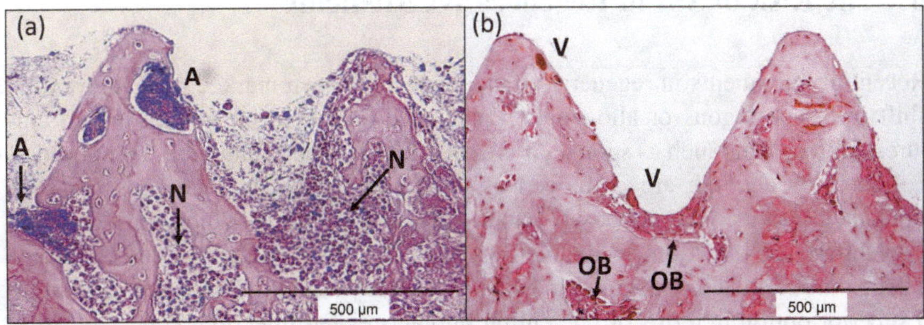

Fig. 10 Histologic section of bone-pin interface for an untreated pin (**a**) and a TiO_2-coated pin (**b**) (Gram staining)
Many abscesses (A) due to bacteria and neutrophils (N) are seen with untreated pins. Few findings of bacteria and inflammation are seen with TiO_2-coated pins, and the screw thread is maintained. Osteoblasts (OB) and vascular connective tissue (V) are seen at the bone-pin interface
Scale bar = 500 µm. Original magnification ×10

from virtually zero to over 50%, and the mean incidence of severe infections requiring pin removal has been reported as 5.8% [27]. When pin site infection occurs with external fixation, various kinds of bacteria including commensal skin bacteria can migrate subcutaneously along the surface of the metal pins toward the internal tissues. As infected pins may lead to loosening, chronic osteomyelitis, and deep infection in the conversion to internal fixation [26], development of antibacterial external fixation pins holds substantial clinical implications. In our experimental system, UV illumination was applied from only one direction for 30 min, but further improvements in photocatalytic activity can be anticipated with multidirectional illumination or multiple rounds of illumination. Since we used a commercially available black light as the UV light source, the intensity was as low as that encountered outdoors, with no exposure to harmful UV-B and UV-C. However, considering the clinical application, negative effects of UV rays on the human eyes and skin cannot be ignored entirely. Moreover, UV-A illumination time to achieve effective bactericidal properties must be minimized. To resolve such problems, numerous studies have been conducted with the aim of clarifying the conditions needed to create highly photocatalytic actions triggered by visible light (fluorescent light or sunlight) through different combinations of crystallites or doping methods [28–30]. Moreover, therapeutic applications in postoperative implant-related SSI have been attempted by providing ultrasonic stimulation from outside the body to elicit photocatalytic bactericidal activity of TiO_2 [31]. TiO_2 does not dissolve or accumulate after implanted in the body, and the photocatalytic activity can be controlled by manipulating the type and amount of illumination. We plan to take maximum advantage of these properties and aim to develop highly antibacterial but harmless biomaterials as an effective measure against implant-related SSI. If greater photocatalytic antibacterial actions can be achieved while maintaining safety, TiO_2 films could contribute to the prevention and treatment of implant-related SSI in the near future.

6 The Risk of SSI in Regenerative Medicine

Recent developments in regenerative medicine have been marked, and an eventual shift from autologous or allogeneic transplantation seems inevitable. In cell cultures, biomaterials such as sponge, nonwoven fabric, sheets, and tubes composed of polymers or ceramic are employed as scaffolding. Regenerative cells are then implanted into the human body together with these scaffolds which carry a risk of implant-related infection. In the Kenzan method [32], regenerative cells or tissue can be transplanted into the human body without any kind of scaffolds, representing a superior option in terms of preventing implant-related infection. However, situations may arise in which a combination of regenerated bone or cartilage tissue is transplanted adjacent to artificial joints or various prosthetic devices. We should therefore understand the characteristics of biomaterials and pay careful attention to implant-related bacterial infections.

7   Check Points

- Artificial biomaterials show disadvantages in terms of bacterial infection.
- Implant-related SSI is extremely refractory, since the bacteria form a biofilm.
- Pathogenic bacteria are commonly staphylococci, particularly *Staphylococcus aureus*.
- Inorganic antibacterial materials using silver, antibiotic agents, and iodine (disinfectant) as well as TiO_2 are the focus of research into countermeasures against implant-related SSI.
- The antibacterial activity of TiO_2 is not species-specific, does not induce drug-resistant strains, and does not prevent osteogenesis.
- An understanding of the risks of bacterial infection with biomaterials is crucial in regenerative medicine.
- The Kenzan method is expected to be superior for preventing implant-related infection in regenerative medicine.

References

1. Hoyle BD, Costerton JW (1991) Bacterial resistance to antibiotics: the role of biofilms. Prog Drug Res 37:91–105
2. Stewart PS, Costerton JW (2001) Antibiotic resistance of bacteria in biofilms. Lancet 358(9276).135–138
3. Nishimura S, Tsurumoto T, Yonekura A et al (2006) Antimicrobial susceptibility of Staphylococcus aureus and Staphylococcus epidermidis biofilms isolated from infected total hip arthroplasty cases. J Orthop Sci 11(1):46–50
4. Phillips CB, Barrett JA, Losina E et al (2003) Incidence rates of dislocation, pulmonary embolism, and deep infection during the first six months after elective total hip replacement. J Bone Joint Surg Am 85(1):20–26
5. Spangehl MJ, Masri BA, O'Connell JX et al (1999) Prospective analysis of preoperative and intraoperative investigations for the diagnosis of infection at the sites of two hundred and two revision total hip arthroplasties. J Bone Joint Surg Am 81(5):672–683
6. Mangram AJ, Horan TC, Pearson ML et al (1999) Guideline for prevention of surgical site infection, 1999. Centers for Disease Control and Prevention (CDC) Hospital Infection Control Practices Advisory Committee. Am J Infect Control 27(2):97–134
7. Kurtz SM, Lau E, Watson H et al (2012) Economic burden of periprosthetic joint infection in the United States. J Arthroplast 27(8 Suppl):61–65. https://doi.org/10.1016/j.arth.2012.02.022
8. Osmon DR, Berbari EF, Berendt AR et al (2013) Executive summary: diagnosis and management of prosthetic joint infection: clinical practice guidelines by the Infectious Diseases Society of America. Clin Infect Dis 56(1):1–10. https://doi.org/10.1093/cid/cis966
9. Todd B (2017) New CDC guideline for the prevention of surgical site infection. Am J Nurs 117(8):17. https://doi.org/10.1097/01.NAJ.0000521963.77728.c0
10. Lentino JR (2003) Prosthetic joint infections: bane of orthopedists, challenge for infectious disease specialists. Clin Infect Dis 36(9):1157–1161
11. Cataldo MA, Petrosillo N, Cipriani M et al (2010) Prosthetic joint infection: recent developments in diagnosis and management. J Infect 61(6):443–448. https://doi.org/10.1016/j.jinf.2010.09.033

12. Kurtz S, Ong K, Lau E et al (2007) Projections of primary and revision hip and knee arthroplasty in the United States from 2005 to 2030. J Bone Joint Surg Am 89(4):780–7851

13. Berrios-Torres SI, Yi SH, Bratzler DW et al (2014) Activity of commonly used antimicrobial prophylaxis regimens against pathogens causing coronary artery bypass graft and arthroplasty surgical site infections in the United States, 2006-2009. Infect Control Hosp Epidemiol 35(3):231–239. https://doi.org/10.1086/675289

14. Prattingerova J, Sarvikivi E, Huotari K et al (2019) Surgical site infections following hip and knee arthroplastic surgery: trends and risk factors of Staphylococcus aureus infections. Infect Control Hosp Epidemiol 40(2):211–213. https://doi.org/10.1017/ice.2018.312

15. Sanderson PJ (1991) Infection in orthopaedic implants. J Hosp Infect 18(Suppl A)):367–375

16. Salgado CD, Dash S, Cantey JR et al (2007) Higher risk of failure of methicillin-resistant Staphylococcus aureus prosthetic joint infections. Clin Orthop Relat Res 461:48–53

17. Charnley J (1972) Postoperative infection after total hip replacement with special reference to air contamination in the operating room. Clin Orthop Relat Res 87:167–187

18. Berbari EF, Hanssen AD, Duffy MC et al (1998) Risk factors for prosthetic joint infection: case-control study. Clin Infect Dis 27(5):1247–1254

19. Fujishima A, Zhang X (2006) Titanium dioxide photocatalysis: present situation and future approaches. C R Chim 9:750–760. https://doi.org/10.1016/j.crci.2005.02.055

20. Sunada K, Kikuchi Y, Hashimoto K et al (1998) Bactericidal and detoxification effects of TiO_2 thin film photocatalysts. Environ Sci Technol 32(5):726–728. https://doi.org/10.1021/es970860o

21. Sawase T, Wennerberg A, Baba K et al (2001) Application of oxygen ion implantation to titanium surfaces: effects on surface characteristics, corrosion resistance, and bone response. Clin Implant Dent R 3(4):221–229. https://doi.org/10.1111/j.1708-8208.2001.tb00144.x

22. Shiraishi K, Koseki H, Tsurumoto T et al (2009) Antibacterial metal implant with a TiO_2-conferred photocatalytic bactericidal effect against *Staphylococcus aureus*. Surf Interface Anal 41(1):17–22. https://doi.org/10.1002/sia.2965

23. Gosheger G, Hardes J, Ahrens H et al (2004) Silver-coated megaendoprostheses in a rabbit model--an analysis of the infection rate and toxicological side effects. Biomaterials 25(24):5547–5556

24. Lucke M, Schmidmaier G, Sadoni S et al (2003) Gentamicin coating of metallic implants reduces implant-related osteomyelitis in rats. Bone 32(5):521–531

25. Koseki H, Asahara T, Shida T et al (2013) Clinical and histomorphometrical study on titanium dioxide-coated external fixation pins. Int J Nanomedicine 8:593–599. https://doi.org/10.2147/IJN.S39201

26. DeJong ES, DeBerardino TM, Brooks DE et al (2001) Antimicrobial efficacy of external fixator pins coated with a lipid stabilized hydroxyapatite/chlorhexidine complex to prevent pin tract infection in a goat model. J Trauma 50(6):1008–1014

27. Green SA, Ripley MJ (1984) Chronic osteomyelitis in pin tracks. J Bone Joint Surg Am 66(7):1092–1098

28. Maness PC, Smolinski S, Blake DM et al (1999) Bactericidal activity of photocatalytic TiO_2 reaction: toward an understanding of its killing mechanism. Appl Environ Microbiol 65(9):4094–4098

29. Li Q, Xie R, Li YW et al (2007) Enhanced visible-light-induced photocatalytic disinfection of E. coli by carbon-sensitized nitrogen-doped titanium oxide. Environ Sci Technol 41(14):5050–5056

30. Koseki H, Shiraishi K, Tsurumoto T et al (2009) Bactericidal performance of photocatalytic titanium dioxide particle mixture under ultraviolet and fluorescent light: an in vitro study. Surf Interface Anal 41(10):771–774. https://doi.org/10.1002/sia.3087

31. Noguchi C, Koseki H (2016) Bactericidal activity of photocatalytic TiO_2 excited by low intensity pulsed ultrasound (LIPUS): an in vitro study. J Orthop Trauma 30(8):S5. https://doi.org/10.1097/01.bot.0000489984.25082.82

32. Verissimo AR, Nakayama K (2018) Scaffold-free biofabrication. 3D Print 431–450

Index

© Springer Nature Switzerland AG 2021
K. Nakayama (ed.), *Kenzan Method for Scaffold-Free Biofabrication*,
https://doi.org/10.1007/978-3-030-58688-1

Index